Also published by
Miller Freeman Publications, Inc.

Volumes in the Sawmill Clinic Library :

Modern Sawmill Techniques Volume 1
First Sawmill Clinic proceedings, Portland, Oregon, February 1973

Modern Sawmill Techniques Volume 2
Second Sawmill Clinic proceedings, New Orleans, Louisiana, November 1973

Modern Sawmill Techniques Volume 3
Third Sawmill Clinic proceedings, Portland, Oregon, February 1974

*Modern Sawmill Techniques Volume 4
Fourth Sawmill Clinic proceedings, New Orleans, Louisiana, November 1974

Volumes in the Plywood Clinic Library :

Modern Plywood Techniques Volume 1
First Plywood Clinic proceedings, Portland, Oregon, February 1974

*Modern Plywood Techniques Volume 2
Second Plywood Clinic proceedings, New Orleans, Louisiana, November 1974

An Introduction to Paper Industry Instrumentation
by John R. Lavigne

Fiber Conservation & Utilization
Proceedings of the Fiber Conservation & Utilization Seminar
Chicago, Illinois, May 1974
edited by Paul D. Van Derveer and Ken E. Lowe

*Guide to the Metric System for the Pulp and Paper Industry
by Ken E. Lowe

*International Glossary of Technical Terms for the Pulp & Paper Industry

Plywood Manufacturing Practices
by Richard F. Baldwin

Timber Cutting Practices (rev. ed.)
by Steve Conway

*Scheduled for 1975 publication

EUROPEAN SAWMILL TECHNIQUES

Contents

Foreword

THIS BOOK is almost certainly the first of its kind to be published in Europe. It differs from existing books on sawmilling in that it contains practical information on mill operation, rather than theoretical principles. It deals with sawmilling as currently practiced in Continental Europe and with many of the new developments being adopted by European sawmill managers to improve sawmill performance, in yield, in production cost reductions, and profitability, rather than for increasing capacity.

The origins of the book lie in an event which itself was a "first" for Continental Europe: The First WORLD WOOD International Sawmill Seminar, which was held in Munich in June 1974 in conjunction with Interforst 74, the Second International Exposition of the Technology of Forestry and the Forest Industries.

The basic objective of the seminar was to provide the sawmill manager with up-to-date information on some of the topics most directly affecting his mill management practices: when to modernize the production line; how to modernize; how to plan sound investment programmes; the question of raw material supply, and how it will affect future industry development and structures; technical details on several new developments in sawing systems; the choice of cutting system for small logs; and papers on several specific cutting systems.

The papers - 14 in all - have been edited for publication in book form along with the spontaneous discussion that took place at the seminar after presentation of each paper. The editing task has proved a demanding one in view of the three different languages involved, and all had to be translated into one language before editing could begin. Every effort has been made to retain the flavour of the speakers' style in the original language; however, some of the texts in the German and French versions of this book which were originally delivered in German or French are translations from versions edited in English.

One section of the book is devoted to sawmilling practices from other regions of the world and from the future. One paper describes mill practices in the Pacific Northwest region of the USA and Canada; two others examine new sawmill technique developments from their regions and assess the suitability of these to the European sawmill industry: one from North America, the other from Scandinavia; a fourth paper describes the world's first wood fibre recovery plant where wood wastes from many different sources are converted into useable fibre.

The level of interest in the seminar was gratifyingly high for those of us at World Wood's European Editorial Headquarters: 450 sawmillers from 40 different countries on five continents attended the event. Simultaneous translation using battery-powered headsets allowed delegates to choose between German, English or French, and discussion went surprisingly unhindered by the momentary pauses needed while a speaker listened to the interpreter's version of the question and a delegate waited for a reply he could understand.

But organizing and holding such an event is much more than a desk-bound operation. Many meetings had to be held to arrive at plans; the views of many experts in the field had to be sought to draw up a list of possibilities for the programme of papers; speakers had to be invited and their presentations discussed. These remarks would be incomplete without an expression of gratitude to those involved: first, to our speakers and especially to Herr Ing. Karl Fronius of the Wood Processing Institute, Rosenheim, for his advice in finalizing the programme; then to Herr Alfons Hammerl, director of the Bavarian Sawmill Association and to the staff of the Munich Fair Authority, for their invaluable assistance in organizing the event; to the interpreters who made the papers comprehensible to all. We wish to thank these people and also those who gave so freely of their advice at the first stages.

Lastly, we hope that this book will serve well in its objective of assisting managers achieve better results with their mills.

Hugh R. Fraser,   World Wood

Programme Organiser   Brussels,

                                                       April 1975

1

Labour costs, log shortages, limitations to investment: factors favouring rationalization

By M. Robert Braun, President and Director,
S.A. Ferdinand Braun et Compagnie, Niederhaslach, by Strasbourg, France

PROBLEMS TO BE SOLVED IN THE OPERATION OF SAWMILLS

Personnel problems

Shortage of labour

Personnel problems are amongst the most important and are the major source of worry in the management of a sawmill.

The most acute problem is the problem of unskilled labour, sawmilling requiring a considerable number of workmen e.g. for sorting the timber, stacking, drying, or loading. Further, where mills have not yet mechanized their timber yards by the installation of plant for debarking, cross-cutting or sorting, it is necessary to employ a large number of unskilled men. The solution consists in employing foreign workers, since sawmills in France now find it practically impossible to recruit indigenous labour. Foreign workers were first engaged from Italy and Spain, but the economic development of these countries attracted migrant workers back home, and today the majority of unskilled workers come from Portugal and Turkey.

The need to employ foreign labour is a severe handicap to the sawmills for a number of reasons: the instability of the working force, their ignorance of language and their lack of experience in the tasks to be performed, all the more serious in that one is often concerned with men who have never worked before. The goodwill of these workers is not in question, and I would like to pay tribute to them. Training them, however, is a very difficult matter for independent firms which nevertheless must undertake large-scale training schemes if they wish to derive profit from employing them.

Why are indigenous workers disinclined to accept unskilled employment in
sawmilling?

The first reason is the poor esteem in which manual work is held by the
European of today. The prestige of the white-collar worker has carried the day
in face of the supposed servitude of the man in dungarees. This spirit is firmly anchored in the present generation, yet it is essential that there should be
a change of attitude, since no economy can function properly if manual workers
are not available.

Working in the open air and often under unfavourable weather conditions is
a second reason.

A third reason lies in the reputation long associated with sawmills for
paying fairly low wages. There has, indeed, been an improvement in recent years,
but much still remains to be done in this.

Shortage of qualified personnel

European sawmills suffer equally from a lack of more highly qualified personnel, in particular of personnel with a thorough knowledge of wood as a substance (for grading or sorting) and from a lack of highly skilled saw doctors,
or fitters for hydraulics or electronics. Indeed, the speed of technological
development calls increasingly for specialists capable of adapting themselves
to pace demanded by the frenzied rate of modernization.

However, it should be possible to overcome the shortage of skilled labour
in France thanks partly to the training provided by schools and training centres
and partly to a law of 16 July 1971 relating to continuous training. This makes
the training obligatory for all firms employing more than ten persons. Each
firm makes a compulsory contribution, which was 0.80% of the total wage bill
in 1972 and 1973, is 1% in 1974, and will reach 2% in 1976. This law, which can
be considered an imposition, is at the same time an opportunity, since each firm
can, and should profit from it by utilising these funds to ensure training for
its own personnel in specialized training centres.

Cost of labour

Added to these various aspects of the shortage of labour we have the cost
of labour, which forms an increasing proportion of overall costs. Thus, as far
as we in the timber industry are concerned, the share in overall expenditure
represented by wages and salaries amounts to 22.55%, that is 1.6% for adminis-
trative staff, 12.50% for personnel employed in the sawmill and 8.45% for social
security payments.

In face of increases in these costs, there have been considerable efforts
to raise the level of productivity. Where before the last war a man would produce 1 m^3 of sawngoods, he would today produce 5 or 6 m^3. In highly mechanized
sawmills today we reckon in general 2 or 3 man h/m^3 of softwood, the range being
wider for hardwoods, where product specification is the determining factor.

Shortage of raw materials

The present situation

In dealing with the second great problem facing the sawmills, the shortage
of raw materials, I can only refer to conditions prevailing in France, particularly in certain of our forest areas. But very similar conditions must prevail

in other countries.

Rapid advances in technology have led most sawmills to undertake modernization and to increase their overall production capacity. Despite a reduction in the number of mills, the overall production of those still in operation has increased. Thus in France ten years ago there were 12,000 sawmills; today there are about 6,500, of which 2,500 may be considered industrial sawmills.

This modernization of the sawmills leads us to ask whether the yield from the forests is able to keep pace. At this moment the answer is in the negative, and I should like to quote the example of Alsace, a region in which I am president of the Association of Sawmill Owners.

The forests of Alsace yield approximately 800,000 m^3 construction timber, but the sawmills in Alsace have a productive capacity of 180,000 m^3 in excess of this figure. This is naturally responsible for round timber being in exceedingly short supply. Consequently its purchase is characterized by cut-throat competition, and the fact that the current practice in France is sale by public auction results in exceedingly high prices, often bearing no relation to the selling price of the sawn timber.

Future prospects — Increasing forest production

But if the forests cannot at present respond to the demand from sawmills, I am of the opinion — an opinion shared by many people engaged in forestry — that the production of the forests could be increased and improved, and this for several reasons :

  - First, more intense silviculture. In France we rely to a considerable extent on natural regeneration of the forests, and the National Forestry Office rarely has recourse to clear felling. Now, as natural regeneration entails a long cycle, it would often be more beneficial to clear the ground completely and then replant.

  - Sawmills dealing with softwoods are less and less inclined to favour large timber, since large timber is synonymus with large defects, large knots, and with hold-ups and interruptions on conveyor chains around the mill. None of the countries where forest production is highly evolved go in for softwood logs of large diameter. I would quote here the examples of some eastern countries, of Canada, Sweden, and Finland. By contrast, in most forest areas in France, whether in the Vosges or the Jura, there is still far too much large timber in the 120- to 180-yr age classes. Now, the era where great beams were required for the navy is past, and the largest size in demand is the 75 x 225-mm plank. To produce such planks, the optimal central diameter is from 30 to 39 cm., which presupposes standing trees with maximum dbh of 50 cm. Higher yields from the forest should be possible by reducing the rotation from 120 to between 70 and 100 yr. I am convinced that the profitability of the forests would then be considerably enhanced, with improvement in roundwood quality and homogeneity. More intensive silviculture should, therefore, be a national duty for every responsible forestry officer, with the aim of overcoming the shortage of round timber.

  - Amongst other actions that might be taken to alleviate the shortage of round timber, I would suggest the reafforestation of all suitable land as well as the planting of certain quick-growing species, such as Douglas fir.

150 m long. For removal and transfer of the sawn timber we used elevators.

This was the first stage, which at the time my colleagues considered rash, not having much faith in my success.

Second stage — 1959

The first stage having been paid for itself very rapidly, we were in a position in the second year to apply ourselves to the second stage, which consisted in introducing a bandmill for large logs.

We began by installing a traditional French-made bandmill with automatic operation and a re-saw and with chains for log feed and transfer of sawn timber. Subsequently we exchanged the original bandmill for a double-cut machine of Canadian design (sawing in both directions), which had a far greater capacity. This modernization forced us, furthermore, to modernize the log yard at the same time, and we installed a 58-m long gantry crane covering the whole yard, allowing us to handle and sort all the round timber.

Third stage — 1969

With the modernization programme proceeding ever faster we became aware in 1969 of the need to undertake a third stage of modernization, which would enable us to remain competitive. This stage involved total mechanization of the log yard, debarking, with cross-cutting and sorting being electronically controlled. Parallel with this development we decided to add a chipper-reducer line for small-diameter logs.

This third stage of modernization proved the most difficult and the most costly, no two mechanized log yards are alike and each one requires its own design. As a result numerous problems arose over the installation and the long process of putting it into operation and I can say from personal experience, that it is easier to construct a number of sawmills than to mechanize a single log yard.

The reasons for automation

Such were the different stages in the modernization of our firm. I would like now to turn to our reasons for making such investments and for undertaking a process of automation. The reasons were very simple: to reduce manufacturing costs to a minimum at all stages of production; in a word, to be competitive, for nobody can evoid the implications of living in a age of frantic competition. I am convinced that if we had not followed the path we chose, our firm would no longer exist.

The financial problems of automation

What financial problems did we have to face in making this capital investment? In the course of the first two stages we took out medium-term loans, the major part of the cost, however, being covered by self-financing measures. For the third phase we undertook a long-term loan from special National Forestry Fund sources available for the modernization of sawmills; the amount of self-finance was still considerable. I must admit that we did not encounter any great financial difficulties, except perhaps a certain distrust on the part of the banking institutions.

Problems for the future

What are the problems of the future? What further measures of rationaliza-

Limitation of investment

A third problem confronting the sawmills is that of investment, when considered in relation to roundwood supply. There are two aspects of this subject to
be distinguished :
1. Investment made with the sole object of reducing labour and consequently
overall costs. Such investment should be undertaken without any hesitation if
profitability is assured.

2. Investment directed towards increasing the volume of production in the sawmills. It is my view that, given the shortage of round timber we are currently experiencing, such investment should cease forthwith. The civil authorities in France are, moreover, aware that they should do this and they no
longer make loans indiscriminately to all sawmills. In practice, there are
special regulations authorizing departments of the Ministry of Agriculture to
permit long-term borrowing from the National Forestry Fund for the modernization of sawmills at a reduced rate of interest, which is a mere 5%. Since
1973 new regulations have been drawn up, and loans may no longer be granted in
regions in which mills are already over-equipped. This is an extremely wise
decision which should be copied by credit institutions in all countries.

Rationalization of production processes

The organizers of this seminar have asked me to treat this subject on the
basis of my own experience, and I propose to express myself very frankly.

The situation before 1957
Up to 1957 my firm operated a sawmill typical of that era using two head
saws, which were framesaws. In addition to these it had a few accessory machines
such as a trimmer, a band re-saw, and a machine for producing firewood.

The logs in those days were brought to the machines in small hand-pushed bo-
gies running on rails; all sawn timber was carried away by the same method. Even
then labour problems were beginning be felt because of the exacting nature of the
work. Manufacturing costs became exorbitant following increases in labour rates
and in social contributions.

This situation convinced me that a big step forward in modernization was
needed. After making several journeys abroad, where I acquired a number of new
ideas, I called upon a designer to draw up a plan for modernizing my old sawmill.

We soon concluded that the complications of modifying the old sawmill were
too great. After due deliberation we decided to proceed with erection of completely new plant on a green field site in view of the importance of having everything in the production line adapted to the whole i.e. both the log yard and the
sawmill itself. By contrast, in an old installation one cannot take account of
all these criteria, and the advantages of new machines is lost unless everything
else fits in.

First stage - 1958
Our plan was to involve several phases, the first of which was already com-
pleted by 1958. First came a fully automatic framesaw line, with secondary machines also automated, i.e. cross-cutting saws, trimmers, chippers for residues,
silos for sawdust, and silos designed to load lorries or railway trucks. The
logs were brought from the log-yard to the sawmill by a feed chain approximately

tion can we apply to our sawmills? For my part, I can think of three from which
we stand most to gain in the direction of rationalization:
- to create the means of providing ourselves with more homogeneous raw materials

- to produce a more standardized range of sawn timber dimensions.

- to offer a greater variety of products.

It would be worth devoting a seminar to each one of these three subjects.
For lack of time I cannot go into greater detail today, but they merit your close consideration.

Discussion

QUESTION: Can you tell us what your rate of conversion is now?

BRAUN: In 1957 it was about 25,000 to 30,000 m^3, and today, for the two mills, it
ranges between 100,000 and 105,000 m^3. However, being situated in the Vosges, a
French Department with substantial forests, the question of conversion quantities
was never so serious as elsewhere. Now, though, we could not expand in the same
way.

GOLFMANN, A.v. Ehrfeld, Austria: What were your reasons for installing framesaws,
then bandmills, then adding a chipper-canter and what differences in labour are
there between the three types of systems?

BRAUN: Exact cost figures are not possible to give, since our accounting system
does not differentiate them. The use of the three systems, however, was because
we thought framesaws gave best results on medium-diameter logs and that the bandmill was best for large-diameter logs and for producing high-quality products.
The third system gave us chips for the pulp industry, and this sector was in a
state of crisis as I pointed out. The idea of several production lines for different sizes of timber works well, and, I might add, it guarantees a good income.

HOLZNER, Maschinenfabrik Esterer AG, Fed. Rep. Germany: Mr. Braun, you said that
while investment to improve productivity and working conditions should be encouraged, investment intended merely to increase capacity should not be encouraged.
How can these two be kept separately?

BRAUN: Yes, I know, it is difficult. But I say this only because in France we
have a very acute shortage of raw material. Sawmills are competing fiercely with
one another, and existence is difficult enough without this. Investment for rationalisation and productivity alone is possible, and I think these can and must
be made. But with a general raw material shortage, I see no room for any outright
expansions in capacity. We have already reached the stage where it seems unavoidable that some mills close down to allow others to survive. I think this is a
difficult question that we will be forced to consider, and it is not easy to find
an answer. Every company is naturally inclined to strive toward its optimum capacity.

2

A modernization programme for a middle-sized sawmill

By Herr Lois Neumayr, Managing Partner,
Sepp Neumayr & Sohn OHG Holzindustrie, Maishofen, Austria.

THE TYPICAL MEDIUM-SIZED SAWMILL IN EUROPE

In our opinion, in Central Europe the term "medium-sized" refers to a saw-mill with an average annual production of 10,000 to 25,000 m^3. In most sawmills of this size the mechanical equipment, beginning at the log yard, comprises a cross-cutting and sorting stage (preceded in some cases by a debarking machine), and a forklift with grab and shovel. Inside the mill there will be a framesaw n and one or two circle saw edgers with built in trimming saws. Slab is converted to chips by a chipper. The sawn timber is sorted on a straight-line or semi-circular green-chain and stacked, then seasoned in the open air in a large storage yard. Many firms already have dry kilns, or are building them or plan-ning to build them, for faster turnover. In the sawn timber storage yard ano-ther front or side lift truck is employed.

The labour force required to operate a sawmill of this size consists of the manager (who in our part of the world is nearly always the proprietor of the sawmill), an office employee for book-keeping, administration and wage account-ing; in the log yard, a man for cross-cutting and sorting and for operating the debarking machine, and a driver for the forklift used for roundwood, who also has the job of transporting chips and trimmings; in the tool shop an expert saw filler and fitter, who is also responsible for the lubrication and maintenan-ce of the saws.

In the mill a framesaw operator and two trimsaw operators (in sawmills which work with two framesaws and one circular trimsaw, there is only one circu-lar saw operator but most often a helper for handling the trimmings, thus in any case three men); at the sawn timber sorting chain, another three men are requi-red, a sorter and two helpers for stacking. In the timber yard there is a lift

truck driver, a foreman and two helpers. For sawmills with their own lorries
or which are a long distance from a good repair shop, it is a great advantage
to have their own workshop and spare parts store, with an experienced foreman
mechanic in charge. The consequent expense is considerable in terms of added
cost per cubic meter, on the other hand the time lost through machines standing
idle for repairs is greatly reduced.

On the basis of this figures a medium-sized sawmill has 16 employees (or 15
men without the proprietor) and will on the average saw 80 m^3 of round timber
per day, which, assuming a working year of 200 days, means an average annual sawing capacity of 16,000 m^3. That is to say, just over 1,000 m^3 per man and year.
This performance naturally depends on the condition of the machines and still
more on the average diameter of the logs. A daily sawing performance of 80 m^3
is not always possible if there is a large proportion of small logs; on the other
hand, with optimum sawing of medium-sized timber with a top end diameter of 30
to 35 cm, performances of up to 130 m^3/day have been achieved with the use of a
normal framesaw and two trimming saws.

Now we come to the definition of "economic organisation of a medium-sized dd
sawmill". In my opinion, it is not so much the productivity of the machines as
the yield and business management of a medium-sized or small sawmill which are
the decisive factors. One can say, in jest, that two kinds of fanatics will be
found among sawmill proprietors: the feed-rate fanatics (maximum performance of
the bandsaw with feed rates up to 12 m/min. and the saw blade speed pushed up to
2.2 m/min, resulting in kerfs which are often 3.8 to 4 mm wide), and the yield
fanatics who restrict the saw blade speed to 1.6 to 1.8 m/min to obtain a kerf of
2.8 to 3 mm and employ a feed rate of only 4 to 5 m/min.

In the case of the "feed-rate fanatic" the average maximum yield is in the
neighbourhood of 65%, whereas the "yield fanatic" gets up to 70% or, with a large
proportion of big logs, often to 73% of sawn wood per cubic meter of round timber.
Naturally these figures cannot be confirmed with 100% accuracy, but such yields
have appeared in exchanges of statistics and been mentioned in conversations with
other sawmill proprietors, especially in South Tyrol and Northern Italy, they
are not quoted only by myself.

In medium-sized and small sawmills the management has greater flexibility
in the purchase of round timber and particularly as regards the sale of the saw
timber. The price for the mostly small lots of round timber is usually settled
personally between the sawmill proprietor and the seller of the timber in a kind
of relationship, and the same applies to delivery and acceptance. This kind of
buying is hardly possible any more in the case of large sawmills, and experience
has shown that, up from a certain size, they have to pay considerable higher prices for their timber, even in the main felling season, and that generally a proper quality control on delivery is not longer possible.

Medium-sized sawmills are in the agreeable position of nearly always being
able to take advantage of market conditions. Due to the smaller quantity of
timber they buy, their financial obligations are not so pressing, so that they
are often able to tide over a slack market for a number of months.

Large sawmills, on account of lack of storage space and their much heavier
financial committments, generally have to sell their timber as it comes fresh
from the saw and, in the case of a stagnant market (wich has not occurred in re-

cent years but will certainly come again) are obliged to sell below the proper
market   price.

I have neither the authority nor any justification for taking a stand
against large sawmills. On the contrary, one has to ask whether, in a few years
from now, an economic existence will be possible for small and medium-sized sawmills in view of the explosion of costs, and whether they will still be able to
complete with the large concerns.

On account of the ownership structure and particularly the mentality of the
proprietors of sawmills in Central Europe, it would probably be extremely difficult to persuade small sawmills to combine to form large concerns, as has been
done in the case of Svenska Cellulosa in Sweden and Metsä-Saimaa Mills in Finland, to mention only two examples.

Returning now to the subject, proposed to me by WORLD WOOD, namely, "The
economic organisation of a medium-sized sawmill", I can only reassert that a sawmill with 15 men and a sawing programme of 15 - 18,000 m^3 per year operates more
economically than a large mill with a capacity of 50 or 100,000 m^3. In this connection I intend the word "economic" to be understood as meaning the greatest
possible yield of saw timber from the valuable raw material, wood, as well as ma-
nipulation and maintenance and the highest possible profit from sales.

COST STRUCTURE FOR A MEDIUM-SIZED SAWMILL

As an annex to the foregoing remarks I have pleasure in including a cost
schedule prepared by the Association of Graduates of the Austrian Sawmill School
at Kuchl (Salzburg), together with a comparison of the return given by a framesaw with normal feed rate of 4 - 6 m/min and the highest rate of 12 m/minute.

Wages, incl. social security contributions	A Sh. 130.- per m^3
Power	16.-
Maintenance	15.-
Borrowed capital	30.-
Administration	30.-
Insurance	10.-
Taxes and other charges	18
General overheads	12.-
Depreciation	30.-
	A Sh. 291.- per m^3

Time allocation at the individual costing centres:

Log yard	0.35 hours/m^3
Saw line	0.85
Sawn timber yard	0.60
Slab	0.25
	2.05 hours/m^3

Higher feed rate versus higher yield

This question is frequently discussed at the present time in sawmill circles. It would be more correct to formulate this question as follows: When should the aim be a higher feed rate and when a higher yield?

The raw material, wood, is particularly subject to fluctuating market conditions and this should be taken into consideration in the planning and management of a sawmill. The sharp rise in price of round timber in the past year lends greater importance to the question of yield. With the use of high-performance machines, the wider kerfs and the idle time which is a consequence of high feed rates result in a loss of about A Sh. 50.-/m^3 (the figure used for "Loss" in example "A" below). Only through accurate accounting is it possible to determine the threshold at which higher performance provides a higher profit. Although these considerations are easy enough to understand, it is not so easy to determine the crucial threshold in advance.

A simple example with assumed figures will help to illustrate the determination of this cost threshold.

Example: A   B

Aim: Highest performance with   Aim: Optimum performance
          highest feed speed (about   with use of 1.8 mm
          12 m/min). Quantity of   saw-blades (max. feed
          timber sawn per year: 28,000 m^3 speed 6 m/min). Quantity
                                                        sawn per year: 14,000 m^3

With a profit margin of A Sh. 100.-/m^3

                  A   B

 28,000 x 100 = 2.800,000.-   14,000 x 100 = 1.400,000.-
-28,000 x   50 = 1.400,000.- (=Loss)
                1.400,000.-

With a profit margin of A Sh. 200.-

                  A   B

 28,000 x 200 = 5.600,000.-   14,000 x 200 = 2.800,000.-
-28,000 x   50 = 1.400,000.- (=Loss)
                4.200,000.-

This example shows that sawmill A, with a profit margin of A Sh. 200.-/m^3 (200 -50 = 150), has A Sh. 1.400,000.- more actual profit than mill B with the same profit margin.

In conclusion, it must be emphasised that the assumed loss depends greatly on sawing technique, the quality of the logs, the skill of the employees, etc. Losses are always calculated on the basis of the kerf width, but every experienced person knows how greatly idle time can be increased through high feed rates (stoppages). Not only rationalisation but, above all, profit-earning capacity will be decisive in the future development of the sawmill industry.

Discussion

LUTZ, S., Anton Lutz KG, Fed. Rep. Germany : I would like to ask on what the sawing programme is based. What type of sawing is involved?

NEUMAYR : As I said in my paper, the mills I am talking about are in mountain
regions. They produce sawngoods both for construction purposes and general uses.
Capacity levels required are not unduly high and we can make a living by working
an 8-hr day.

HERR FRONIUS, Programme Director : I would like to add something here. Sawmill
performance has generally been judged in recent years by one figure alone :
paid man-hours per cubic meter sawn; alternatively, this figure has been given
in terms of cubic meters per man per shift per year. Herr Neumayr gave the
figure as 2.05 man-hours per cubic meter, or approximately 1,000 m^3/man/yr.
Though these figures have not been used so very frequently here, I must emphasise
that they are not representative any longer, because of the steep rises in raw
material costs. What we do need to know is the net return to the sawmill per
unit of production. We should avoid undue emphasis on man-hour per cubic meter
figures, because they can be misleading.

3

Use of chippers and reducers in high-production sawmilling

By Herr Ing. grad. Karl Fronius, Director,
Institute for Wood Processing, Rosenheim, Fed. Rep. Germany.

Introduction

The problems of the economic cutting of round logs with conventional machines have hitherto arisen because costs could not be covered in the case of small logs and in the case of large logs irregularities in stems often led to serious losses in output.

Two comparatively recent concepts in the Central European sawmill industry, reducers and canters, have opened up new possibilities for improving the production of sawnwood and increasing sawmill profitability.

Reducers

Reducing is understood as a fully mechanised preparation of the logs for sawing. As is well known, the performance of framesaws is greatly impaired by swellings at the butt and elsewhere, knots, and also irregularities in growth such as crookedness, pronounced taper, and through the saws being fouled by the slabs or seizing in the logs; also by inadequate seating of the log and loss of time through aligning the logs in the framesaw.

Reduction takes place between debarking and sawing. A distinction is made between:

a) cylinder reducing by cylinder reducers

b) flat reducing by canters.

Both types of machine perform a milling operation in which all irregularities of the log that could impede sawing are removed. The material removed by the milling cutters has the shape and size of standard chips and can therefore be marketed.

Cylinder reducing

Mode of operation - The object of this technique is primarily to remove large butt swells and other growths that could impede the proper functioning of frame or circular saws, and at the same time to reduce the irregular diameter of the log to a suitable dimension for the framesaw. The round reducing machine operates in the same way as a round bar milling machine but has a considerably larger infeed aperture. At their critical points the logs are given a cylindrical shape, which greatly facilitates their handling at the saws reducing lost time to a minimum. The quantity of chips produced is relatively small and not of very high quality, since it contains a large amount of ingrowing bark, particularly at the butt swell.

Machines - Cylinder-reducing machines are manufactured in various types:

a) with a fixed rotor aperture (400 mm Kockum)
b) with an adjustable rotor aperture (200 to 400 mm Kockum)
c) with revolving milling cutters in the debarking rotor (Braun)

Cylinder-reducing machines with a fixed rotor aperture ensure that all other machines can be set up for a definite limit of log diameter.

Machines with an adjustable rotor aperture are used in sawmills in which the cants are sorted before sawing.

These cylinder-reducers are generally coupled direct to the debarking machines. In large sawmills employing framesaws, two reducers are frequently operated in tandem in order to avoid trouble with slabs as far as possible.

The debarking machine also controls the feed rate, which depends on the diameter of the log (22.5 to 70.5 m/min).

At the present time tests are being carried out with a hollow rotor debarking machine which debarks and round-reduces in one operation, the diameter of the aperture being variable.

Advantages of cylinder reducers.

1. A constant feed rate to all machines, since practically no hold-ups are to be expected. Increase of production up to 20%.

2. Limitation of the width of the framesaw, thus enabling smaller but faster framesaws to be used, which also results in an increase of production of about 10%.

3. Automation of the feed to the framesaw and full mecanization of the trimming operation. This effects a considerable saving of highly-paid labour.

4. rationalization of the whole production line (machines and conveyors) through a maximum limit of log diameter.

5. Decrease in the quantity of slabs, resulting in fewer notches in the side boards and thus less need for subsequent operations.

6. Decrease in the risk of accidents at the main sawing machines.

Flat reduction

Mode of operation - Flat reduction means the production of parallel flat
surfaces on the logs before sawing. For this purpose face-cutting machines,
also known as canters, are employed. These machines can be used for two different purposes, either

a) flat reduction or

b) the production of saw timber.

Advantages - The idea behind flat reduction is to chip off the wane or slabs
that otherwise are produced when round timber is sawn. This is done either before or after sawing. Chipping before sawing brings the following advantages:

1. Log breakdown is practically trouble-free and also labour-saving whatever
breakdown system is used.

2. Sawn material is easier to remove and handle after sawing.

3. Trouble-free collection and removal of the slab material in the form of
chips.

4. Kerf losses on sideyield are eliminated. Instead, commercially useable
chips or, in part, sawntimber are produced and result in a real improvement
of the yield.

Reducer bandmills - Flat reducing has recently been supplemented and extended by the use of reducer bandmills. These are a combination of a chipper
and a twin bandsaw. The chipper first profiles the log and then trims the
rounded edges, after which the twin bandmill makes two parallel cuts.

All the advantages of the systems already discussed are provided by this
new reducing technique and, in addition, there is a considerable gain in the
utilizable diameter of the log, since, besides flat reduction, a board is produced on each side of the log under optimum cutting conditions (feed and yield).

Profile chipping techniques

Mode of operation

A series of factors has opened up new interest in the fully automatic
processing of roundwood and in the marketing of the product especially in the
case of small logs (12 to 22 cm dia). These are: difficulties in the sale of
inferior quality sawntimber, constantly increasing operating costs, and the ex-
haustion of the possibilities of rationalizing conventional sawing methods by
mecanization.

With profile chipping the logs are not initially cross-cut and then reduced
but are immediately milled or chipped in order to eliminate the sawdust loss of
10 to 12% that results from sawing. As is well known, there is no market, or
only a very poor market, for sawdust.

The method used by European manufacturers of sawmill machinery (Gebr. Linck
German Federal Republic; Kockum, Sweden) is to process two sides of the log in
one operation and the other two sides in a second operation. A Canadian firm
(Chip-N-Saw), on the other hand has designed a cantering machine which processes
the log on four sides simultaneously, resulting in the production of a squared

cant. The first method has the advantage that crooked stems can be better and
more easily aligned for profiling, thus reducing waste. Important here is the
centering device ahead of the canter.

(Note: American techniques are not dealt with here in greater detail, since they
are the subject of a separate paper)

Yield

Sawntimber yield depends on a number of factors such as: kind of wood,
log taper, log length, crookedness of the logs, and wane.

The following figures can be given as an average guide (expressed as percentages of round timber before machining)

Sawntimber	46 – 56%	
Chips	38 – 46%	
Fine wastes	4 – 8%	(without any sawing)

Marketing the products

Sawntimber – Since profile chipping is primarily applicable only to small
logs, the variety of dimensions that can be sawn is very limited and there is
a danger of flooding the market with certain sizes. The following sizes can be
produced or present themselves:

Squared timber in the range from 6 x 6 to 16 x 16 cm maximum.
Half-struts (by division of squared timber)
Chamfered narrow boards in the widths 8 to 14 cm.

Possible users are prefabricated wooden house producers and manufacturers
of glue-laminated products, underground construction contractors, and carpentry
workshops.

Chips and shaving specifications – Depending on the machine used, the following kinds of waste are produced:
Flat chips 0.2 to 0.4 mm in thickness for the chipboard industry and chips in
different lengths for the fibreboard and cellulose industries. It should be
noted that the chipboard industry is also a user of hacked chips.

Fine waste – Depending on the make of machine used and the size of the cut,
4 to 8% of fine waste (fraction 5) is produced. It is sometimes rejected by
the cellulose industry, but mostly accepted by the chipboard industry, according
to the state of the market.

Infeed opening and depth of working

With logs of medium size up to a maximum diameter of 40 cm the infeed opening should be wide enough to allow face-cutting to be carried out.

With small logs the depth of working should be at least large enough to
ensure that sawnwood of marketable cross-section can be produced from the cant.
Roundwood unsuitable for the production of sawntimber should be converted entirely into chips in one operation.

Profile chippers currently available on the market have admittance widths
up to 400 mm and depths of working of 2 x 100 up to 2 x 190 mm.

Production capacities

The sawntimber production capacity of a canter installation depends on
seven factors. These are:

       a) Number of faces worked at a time (two or four)
       b) Feed speed (22.5 up to 60 m/min)
       c) Log diameter, average (12 to 22 cm)
       d) Sorted/unsorted logs (70%/50% of maximum theoretical machine
                    utilization respectively).
       e) Average length of the log (1.0 to 7.0 m)
       f) Residue specifications (flat chips or hacked chips).
       g) Infeed and log alignment design (determines log piece count).

Another factor is whether a single profile chipper is used or two in tandem. If
the output of one canter is put at 100%, then that of two in tandem will be
about 230%.

Surface quality on sawntimber

Surface quality is a very important criterion in assessing profile chippers.
In general, it would be assumed that the use of cutters would result in a better
surface than that given by sawing. But that is not always the case, since the
feed rate, the growth area of the timber and its degree of dryness are deciding
factors.

Very dry logs are particularly liable to tear at the knots left by branches.
This can be prevented to a great extent by the use of cutter teeth segments.
These do tend to increase the proportion of small chips, but sometimes a com-
promise is unavoidable. At high feed rates the cutter teeth leave marks. These
difficulties can be overcome by passing the material through a coupled planing
machine, which ensures a good surface and also helps to maintain uniform dimensions.

In some cases sawntimber is produced with an allowance for shrinkage and
processing, so that it can be kiln-dried and then planed before delivery. In
this way surface faults are mostly eliminated and the dimensions of the timber,
which are somewhat changed after the drying process, can be calibrated again.
This procedure, however, entails a certain loss of yield.

Chipping edgers

The profile chipping concept can also be applied to edging, in place of
a twin circle saw edger. With a chipping edger, edgings are converted directly
to chips by rotary chippers and at high feed speed: over 100 m/min. This system offers the following advantages:

1. Trouble-free collection and removal of chips.
2. No danger of recoil.
3. Elimination of gate and conveyor for edgings, allowing a sawhouse some
   7 m shorter.

Advantages of profile chipping techniques

There are many advantages in comparison with conventional methods of cutting timber. Among these can be mentioned:

    1. Production of only two varieties of goods (the main product and

chips), simplifying the question of storage
2. High productivity (up to 4,000 m^3 per man and year)
3. Less investment for auxiliary machines, conveyers and buildings
4. Saving of power, since milling is carried out immediately, without preliminary sawing
5. An increased yield of 4 - 5%, since there is no or only a limited production of sawdust
6. A considerable saving in cutting costs compared to conventional methods of cutting (particularly labour costs).

Combinations of machines

The chipper is extremely versatile in its applications, especially in combination with other sawing machines. The following are the basic combinations (figures for production are given in terms of unsorted and sorted round timber, respectively):

1. One profile chipper, possibly with an associated retracting circle ripping saw for cutting small logs in only two cuts into single and double squared timbers. Annual output 31,000 - 43,000 m^3.
2. Two profile chippers in tandem with a straight-line feed to the second unit, which could be fitted with a retracting ripping circular saw for the same sawing programme as 1 above. Annual output 71,000 - 99,000 m^3.
3. One profile chipper followed by a two-blade circular saw or twin bandmill and a retracting ripping circular saw with a merry-go-round for sawing logs of medium size. At each pass of a log the 2-blade saw produces a board from each side, which are then passed through a chipping edger. The ripping circular saw serves the same purpose as in 1. Annual output 49,000 - 69,000 m^3.
4. Two profile chippers in tandem with straight-line transfer, followed by a two-blade circular saw or twin bandsaw and a retracting ripping for cutting logs of medium size. When the log passes through the second canter a board is produced from each side, these are then passed through a chipping edger. The ripping circular saw serves the same purpose as in 1. Annual output 113,000 - 158,000 m^3.
5. One profile chipper possibly with retracting ripping circular saw with merry-go-round, followed by a framesaw and a chipping edger. The profile chipper, as under 1, is used for the first and second cut to convert small logs into squared timber or for the preliminary shaping of medium-sized logs for final cutting in the framesaw. Sideyield goes to the chipping edger. Annual output 40,000 - 56,000 m^3.
6. One profile chipper followed by two framesaws and two chipping edgers for logs of medium size. The profile chipper is used exclusively for initial breakdown and the two framesaws for the final sawing of the timber. Annual output 40,000 - 56,000 m^3.
   Note: Combinations 5 and 6 can be still further varied for small and medium sized logs.
7. A reducer bandmill with merry-go-round followed by a ripping bandsaw and a chipping edger for the production of squared timber, planks and boards from medium-sized logs. Annual output 50,000 - 70,000 m^3.
8. Two reducer bandmills, a twin bandmill, and a ripping bandmill in

line, followed by two chipping edgers for the same programme as under 7.
Annual output 115,000 - 161,000 m^3.
<u>Note</u>: The calculation of the annual output is based on the processing of
unsorted round timber, for sorted logs
add 40%
Average diameter: a) small logs 17 cm
                    b) medium sized logs 24 cm.
Total annual working hours: 1,800 hours.
Operation in tandem: 130% additional output.
Gross feed rate: a) Small logs 50 m/min.
                   b) Medium size logs 40 m/min.

Summary

Use of chippers is a prerequisite for full automation of sawntimber production. It enables new standards of efficiency in sawing to be attained with an
accompanying improvement of yield and productivity.

Reducing with chippers as an alternative to sawing was originally designed
for the rational processing of small logs. In the course of its further development, however, it has already been applied partially to the processing of medium
sized round timber.

The high efficiency of modern reduction technique is illustrated by the
fact that three lines using small logs with one employee only are already in
production. The operator is able to control the whole process from the delivery
of the logs to the production of the finished sawntimber with an annual output
of 50,000 m^3 on the basis of normal working hours.

But any modern technical process is incomplete if, in the end, the marketed
product does not show a profit after the cost of raw material and sawing has
been covered. Our processing machinery, must be very flexible in its applications. For, let us remember, what was the relationship five years ago between
a ton of side boards and a ton of chips, what was it last year and what is it
today. Rational processing in the sawmill industry is, therefore, not the only
decisive factor; the primary factor is constituted by the market to which we
sell our products. This means that economics always comes before technology.

Discussion

NEIDHARDT, F., Firma Fritz Neidhardt, Fed. Rep. Germany: Can you say to what
extent the figures you have given are applicable to the larger diameter of log
we encounter here as compared to those Scandinavian diameters you quoted?

FRONIUS: This is answered in my paper but I could add a little. Feed rates for
a framesaw line equipped with a rotary reducer are within the normal range of
debarking feed rates in central european sawmills, that is to say, 25 to 35
m/min. However, it is best that you obtain more precise details from the
company Braun in Augsburg, which is not far from here.

4

The fully automated sawmill: its operation and economics

By Professor Bertil Thunell, Swedish Forest Products Research Laboratory, Stockholm, Sweden.

Seen in a broader context, sawmilling is an industry with a very wide technical spectrum, ranging from rudimentary, primitive sawmills to highly sophisticated installations, from very small sawmills to sawmills with a very high output. This diversity arises from differences in the conditions applying in each particular case, and one is not entitled to condemn or recommend any particular type in general terms. One statement is possible: the relatively simple sawmill equipment of yesterday is now being supplemented by equipment in which full justice is done to the scope of modern techniques.

Some idea of developments to date can be derived from figure 1, which shows how the average number of man hours per cubic meter of sawntimber has declined in Sweden during the last ten years since 1963. The statistics reveal that there is still quite a wide dispersion around this figure and that the sawmills with the lowest values are appreciably below the level indicated by the graph.

Here we shall be dealing with the group of sawmills where mechanization is most advanced and where operations are controlled by electronics and servo-systems as far as possible.

By way of a background description, however, something should first be said concerning the general cost situation in the manufacture of sawntimber products. For many reasons, it is natural for the figures quoted in this connection to refer to average conditions in Scandinavia. The figures have been rounded off.

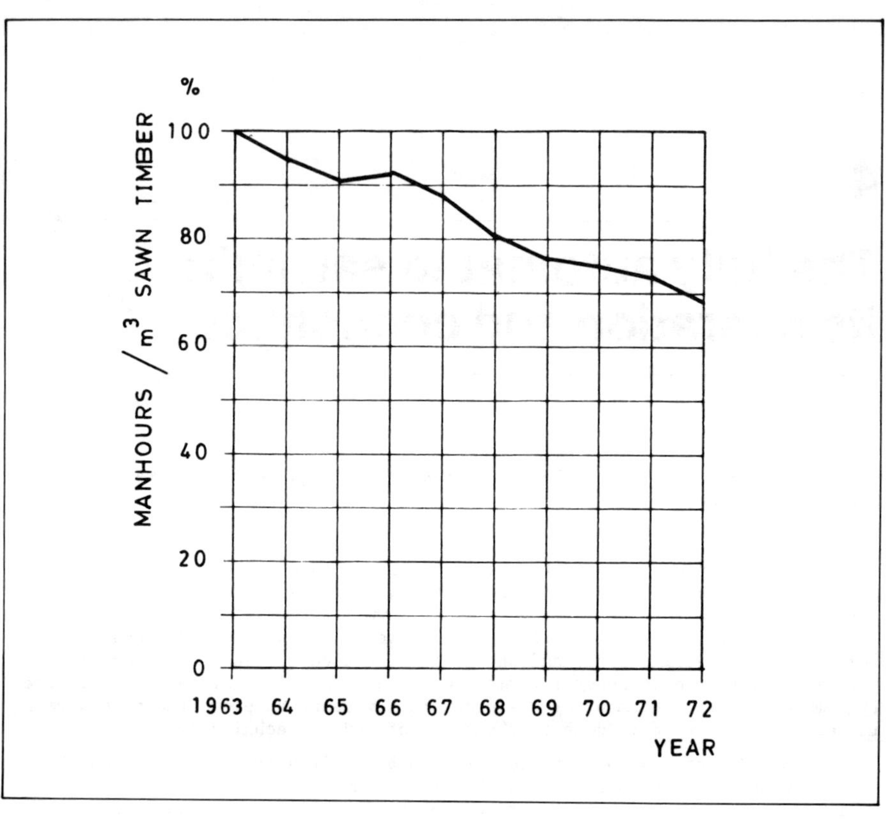

Fig. 1 - Development of the average number of man
hours per m³ of sawn timber.

Table 1. Average cost picture

The costs involved in the production of one cubic meter sawntimber break down as follows :

Labour costs	6 %
Capital costs	9 %
Miscellaneous production costs	8 %
Raw material costs	77 %

The actual volume return in finished products, stereometrically calculated, is put at approximately 55%.

Between 18% and 20% of raw material costs are recovered by selling the residuals in the form of chips, shavings and sawdust.

These figures show that the input of labour today is low and that the labour costs are a relatively small part. A further reduction of labour input, e.g. by 20%, will reduce production costs by about 1% but may result in a corresponding rise in investment costs. The decisive factor is the effect on raw material costs.

The picture of costs given above can be broken down still further and costs distributed between the various stages of processing in the sawmill. The following will then be obtained:

Table 2. Production costs

Production costs in % divided between	Timber handling	Sawing	Drying	Dry timber handling and storage
Labour 6%	11%	45%	4%	40%
Capital 9%	15%	35%	30%	20%
Miscellaneous production 8%	20%	40%	10%	30%

This table shows that the operations which can be said to process and refine the timber, i.e. sawing and drying, account for only about half the cost. Timber handling and dry timber handling plus storage are clearly in need of further development on the basis of complete systematic studies. The relatively low cost of processing justifies a development of processing and handling techniques primarily on condition that this will give better utilisation of the raw material.

If matters are viewed in these terms, the following aims must govern the sawing of every log:

1. Optimum sawing pattern must be used for each individual log
2. Saw kerfs must be as small as possible
3. Sawn product must be correctly dimensioned
4. The surface profile of the sawn product must be adapted as closely as possible to its subsequent use
5. Drying must not involve degrading

The special implications of this for processing can be summarised under the following heads.

Machinery

   1. Precision in machine setting
   2. Sound function of mechanical components
   3. Sufficient flexibility
   4. Machine stability
   5. Saw tool stability
   6. Adequacy of cutting capacity

Log positioning and saw infeed.

   7. Correct positioning at infeed
   8. Correct movement along the log length

Simple and obvious as these requirements may seem today, they have not been easy to satisfy. In some cases a great deal of work still remains to be done before adequate solutions can be achieved. The first four items are mainly the responsibility of the machine designer, acting within the framework of available knowledge and techniques. Items 5 and 6 have long been a subject of research and development in many quarters, while items 7 and 8 have only recently attracted the attention they deserve.

Only when the entire system, man-machine-material is studied on the basis of ergonomics, can the right paths for an efficient system by indicated.

The reduction in the number of working hours per cubic metre of output referred to above has, among other things, been achieved by means of increased feed velocities. This in turn has made it increasingly difficult for the operator to do a proper job of work. Automatisation has then served to lighten his burden so as to confine the human contribution to the operational stages at which it has been most necessary. This applies above all to the control of the conversion process at points where, for technical or economic reasons, automatic control is not suitable.

Since no complete mathematical quality descriptions have yet been applied to sawntimber in industry and since there are still no automatic sensing and selecting systems for all possible timber defects with a bearing on quality assessment, any assessment must be finally determined or supplemented by human intervention. This applies to higher quality, larger diameter logs. Where small logs n are concerned, it is economically justifiable to neglect some of these assessments and to design more highly automated infeeds and to saw solely with regard to geometrical shape.

Summing up the above observations concerning the automatisation of sawmills we obtain the two extremes of utility of the automated sawmill where:

   1. Automation gives good raw material utilisation.
   2. Automation allows production of a bulk commodity from a cheaper raw material at a very high output rate and with little or no manual input.

Between those two extremes many stages and combinations will easily be found in the industry.

One important factor is that a high level of automation be combined with the demand that the production is large enough with regard to costs, which usually have a high fixed component.

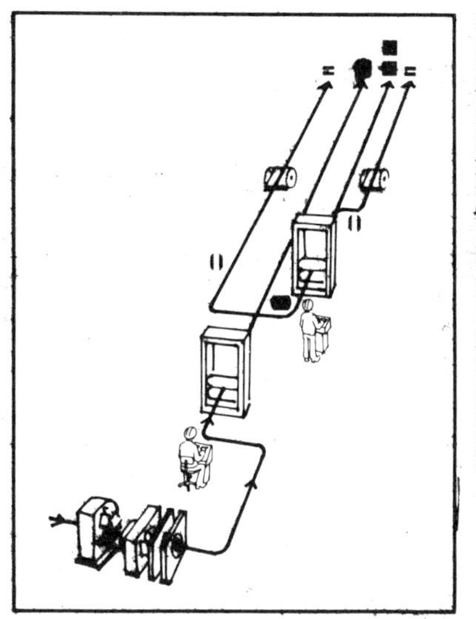

Fig. 2 - Framesaw line with edgers.

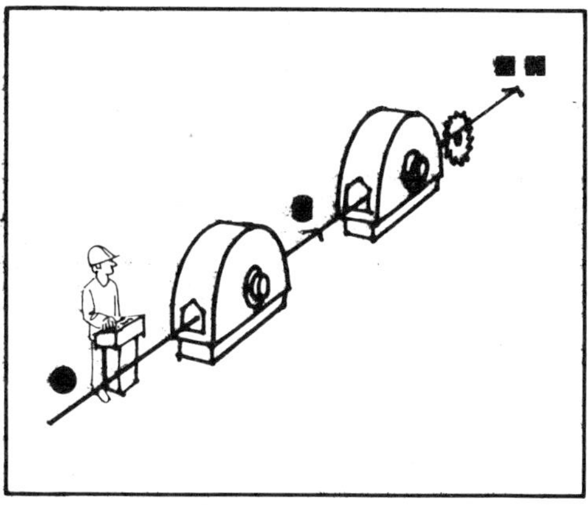

Fig. 3 - Line with two reducers and a resaw.

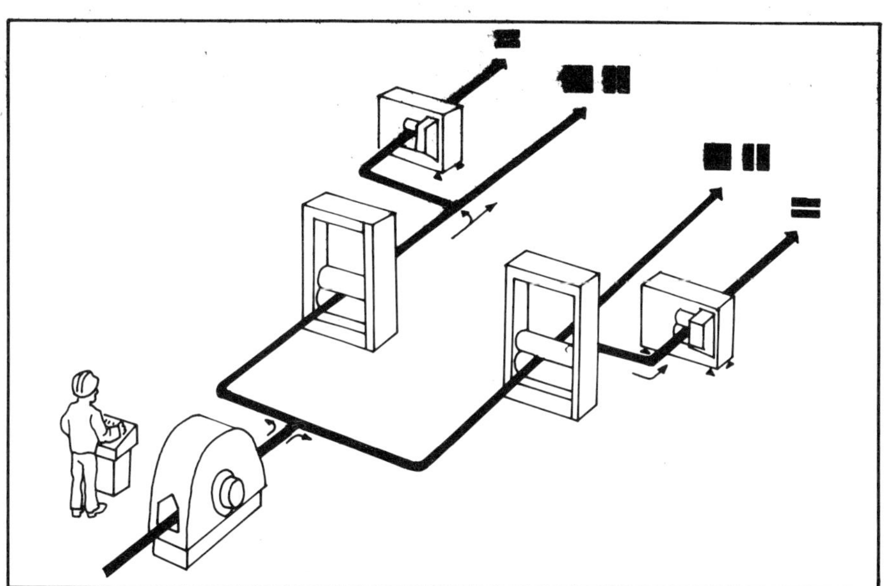

Fig. 4 - Line with a reducer and two framesaws for dealing.

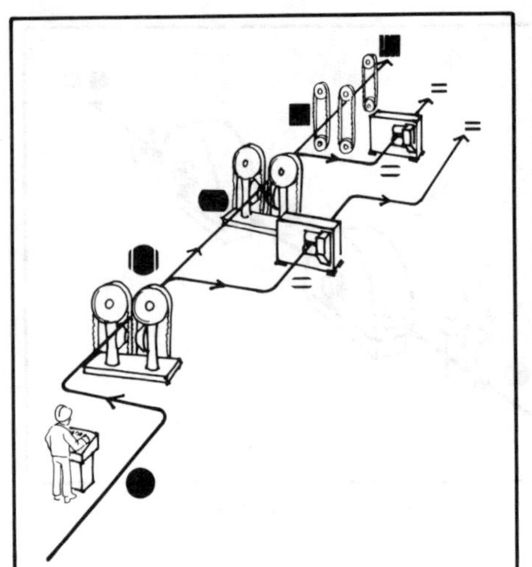
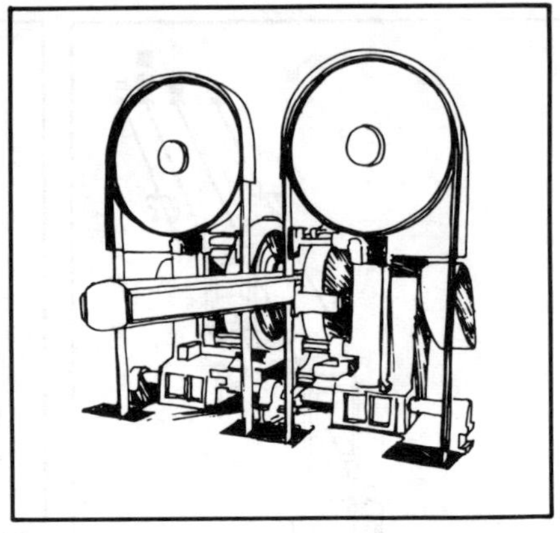

Fig. 6 - Reducer bandsaw.

Fig. 5 - Reducer bandsaw line.

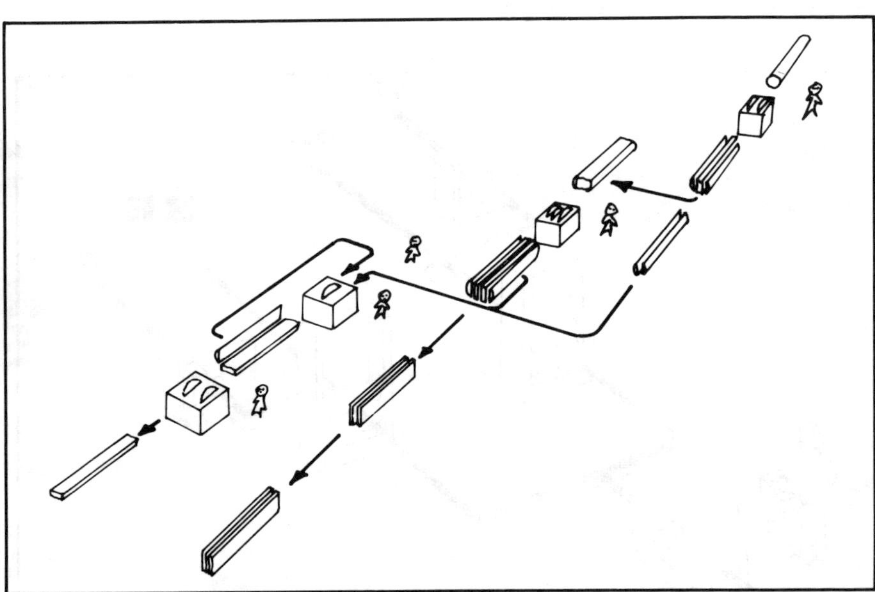

Fig. 7 - Circular sawmill, system 25.

Examples of the two cases cited are provided by a number of recently constructed sawmills in Scandinavia. These include mills specifically of type 1 or
type 2 and also a number of mills which have the two types in separate production
lines.

The frame line, as in figure 2, with a cant and deal frame together with
an edger for the side boards, is operated by one man who controls infeed and the
first framesaw. This type of layout is used with various machinery units for both
thick and thinner timber. The pure framesaw line as shown here is not the commonest
form adopted for new installations in Sweden today. Where small diameter logs are n
concerned it has been superseded by chipper-canters combined with circular, band
or framesaw for dealing. This type of arrangement is illustrated in figure 3.

Various combinations of reducers, circular, frame and band saws are used
for middle-sized timber. Thus figure 4 shows an arrangement with a reducer for
producing cants and two frame saws for dealing.

Due above all to its high output capacity and flexibility, the reducer
bandmill, combining reducers and bandsaws in a single unit, has attracted a considerable amount of interest. High output capacity and flexibility are its main
advantages. An arrangement of this kind is shown in figure 5 and figure 6.

All these layouts give high output capacity and are therefore characteristic of the larger units in the Scandinavian sawmilling industry.

In Scandinavia's numerous medium sized sawmills, however, the circular
saw has played a leading part and still does. This technique has also been developed as regards both machinery design and layout. Consequently a circle saw-
equipped mill is completely different from before. Both mechanisation and automatisation have progressed a great deal.   Figures 7 and 8 show the layout of
modern sawmills of this kind, with multi-blade circular saws, and a sawmill interior. Here too a modern combination of infeed system and saw units can result in
an efficient sawmill even with small outputs. The design also allows the sawmill
to increase its capacity should the need arise. The next three show some expansion stages on these lines as drafted by Ackerfeldt. Sawmills of this kind should
be of particular interest to the central European sawmilling industry.

Here the operator sits at a control panel and positions the log with the
aid of the turning device, aligning shadow lines on the top and end face. At the
same time the aligner measures the diameter of the top end, and this measurement
can if so desired be fed to a small, simple computer, to determine the sawing programme and even to control the sawing process automatically. The log is kept in
its correct position by fingers on the aligner.

When sawing is about to begin, a rear arm is lowered and moved forward
against the end of the log, moving the log forward until its butt end contacts
the front arm.

Before the log starts to move forward, the blades in the double saw are
set to saw out a slab from each side. The arms have retained their hold on the
log and return with it immediately until the front arm has emerged from the saw,
whereupon they stop while the blades are set to the dimension required for the
next pass.

If no additional parallel cuts are to be sawn, the arms now deposit the
cant on the table. The operator makes sure that the block is turned through 90°

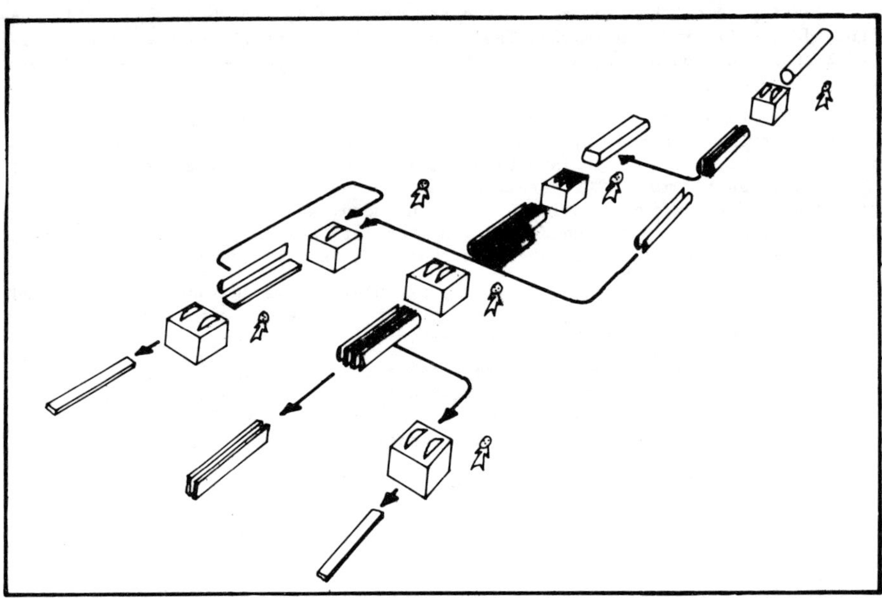

Fig. 8 - Circular sawmill, system 35.

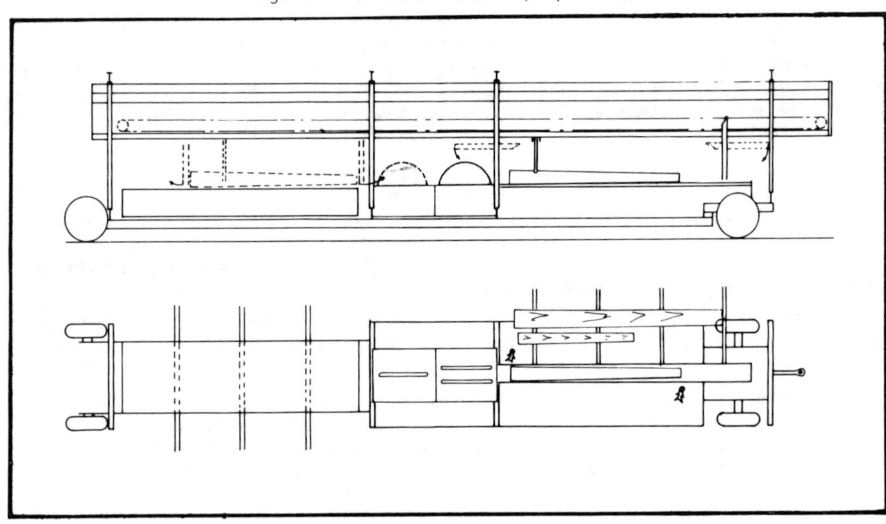

Fig. 9 - Modulac sawmill, module 1, mobile.

so as to rest on one of the sawn sides and the log is then aligned in the manner described above and the sawing starts again.

Module 1 is the smallest unit. Figure 9 shows a mobile version. The intention has been to provide a very cheap sawing installation for conditions where a low level of investment is of great importance, where labour is readily available and where good capacity is, nonetheless, sought. Of course, module 1 will require even less investment if it is made stationary. Sawing is managed by two men.

Module 2 is a sawmill with edgers, a debarking machine and a chipper as well as a sorting unit for the sawntimber. It requires four men. Output approximately 6000 m^3/yr (Fig. 10).

The sole difference between modules 2 and 3 (Figs. 10 and 11) is that in the latter case the centre saw is shifted one log length away from the circular saw. This makes it possible for the cants to return round the saw to the infeed side instead of accompanying the feeder arms when they are retracted. This saves a considerable amount of time because a new log or block can then be fed at an earlier stage. The main additions are an automatic split saw replacing the central saw, the necessary conveyors and the steering devices in the roller conveyor leading the centre piece back.

It is evident from the above that the sawmilling industry today stands to benefit from a high level of automation even when small units of production are concerned. It also seems reasonable to say that selection must be based on the character of the raw materials concerned and on current socio-economic circumstances.

A survey of where electronics and servosystems can be used in the sawmill production is given in figure 12.

In doing this, the production process is divided into ordinary steps:

1. Roundwood handling with dimensional sorting of logs.
2. Sawing
3. Dimensional sorting on the greenchain
4. Stacking
5. Drying
6. Unstacking and trimming
7. Packaging

All these steps can be automatised to a greater or lesser extent. This automation is often combined with necessary reporting and recording of results for production control and stock accounting.

The difficulties confronting further development come about with variability in the raw material, in terms of both dimensions and quality. Further difficulties arise with the impossibility of describing the quality requirements of the end product in simple quantifiable terms: these requirements can even be fundamentally different with different applications, which are often unknown at the time of production. Another difficulty is that deviations of shape and dimensions of the sawntimber can impose exacting demands on sensing devices, controls and mechanical systems.

Consequently the complexity of the systems required increases rapidly with increasing applications in the mills, as does their sensitivity to disturbances.

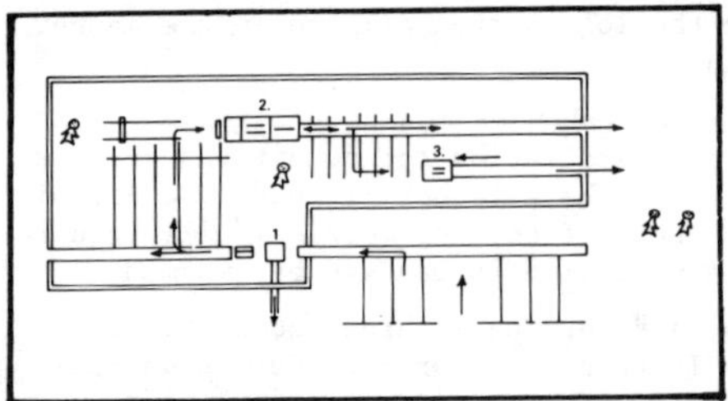

Fig. 10 - Modulac sawmill, module 2, for a production of 5,400-6,700 m^3/yr.

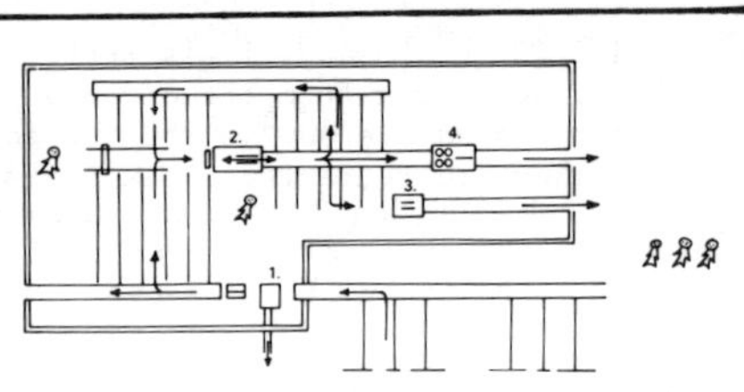

Fig. 11 - Modulac sawmill, module 3, for a production of 6,800-8,300 m^3/yr.

Fig. 12 - The "automated" sawmill.

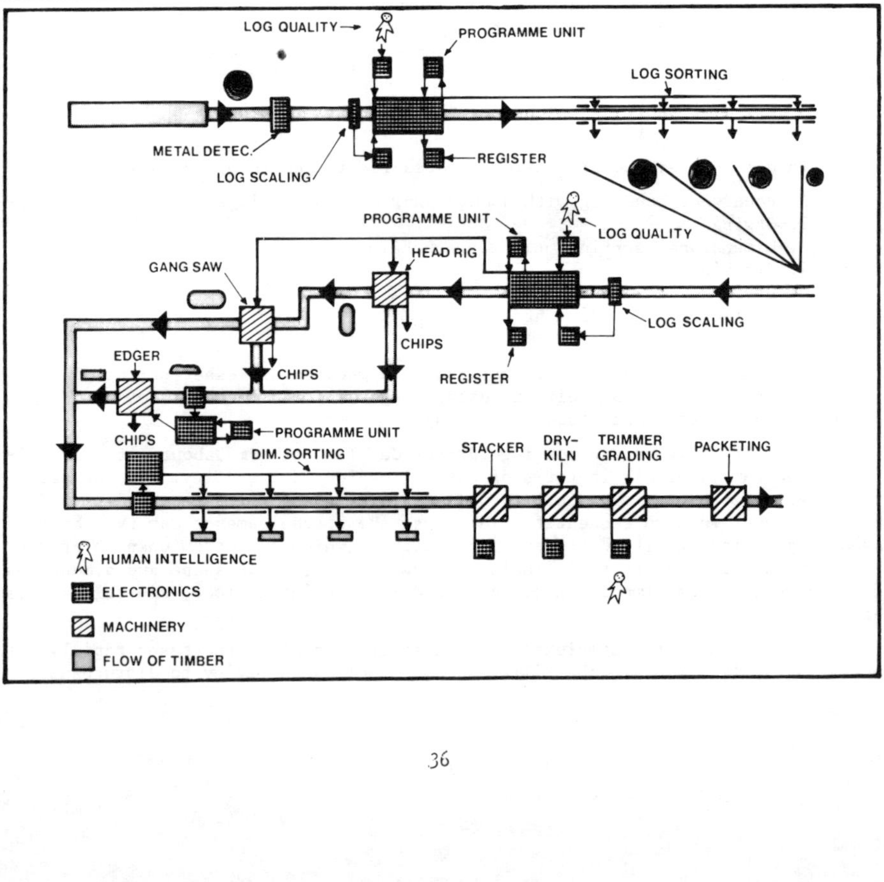

A great deal of expenditure is required on the necessary preventive maintenance in order to keep breakdown costs to a reasonable level.

The likeliest conclusion to be drawn from the above is therefore that further automatisation can hardly be justified on grounds of reduced labour costs but may to some extent be justified by the possibility of an increased rate of production and, above all, a better utilisation of the raw material.

Discussion

GROSSMANN, J., A.v. Ehrfeld, Austria: Professor Thunell, I would like to ask you what annual production capacity is needed for the economic operation of a reducer bandsaw.

THUNELL: I would like to refer this question to Mr Morgenstern, who is from the Kockum Company in Sweden.

MORGENSTERN: Exact figures are difficult to give, but our experience indicates that even small mills find economic advantages in using these machines. But with variations in raw material, yield, and differences in markets, no general rule can be made on this and each case has to be calculated with its own data.

QUESTION: What interests me is the time required for this entire automated line to be stopped for a change of tools. For what duration do the bandmills and chippers operate and what length of stoppage is needed to change tools?

THUNELL: In Sweden, work is divided into 4-hr periods. This provides an opportunity for sawsteel to be changed every 4-hr. Exactly how long it takes to make this change I am not sure, but I think something in the order of about ten minutes would be close to it.

5

Determining the most economical cutting system for small-logs: gangsaw, bandsaw, chipping headrig, or circular saw

Panel discussion

Lead by Herr Karl Fronius and Senior Forest Officer Wolfgang Steuer, Forest Administration Headquarters, Munich, Fed. Rep. Germany.

FRONIUS: The term "small logs" is understood to refer to coniferous stems of any length, with an average diameter of 12 to 22 cm and suitable for the production of, among other things, certain dimensions and grades of sawntimber, sawn boards and single and double cut scantlings.

The market for small logs is of increasing importance. A larger supply of small logs can be expected in the future due to reduced rotations and, especially, as a result of the extensive reafforestation carried out after the second world war.

In recent years the machine suppliers have introduced a number of new types of machine for the economic processing of small logs. Among these can be mentioned high-speed frame or gang saws, multi-blade circular saws, profile chippers or canters, operating on two or four surfaces, and combinations of these machines arranged in production lines.

The cost of raw materials, of the mill and its operation, the output, yield and the market price of the products are all factors that combine to produce a complex problem in relation to the economic use of one type of machine or the other.

This problem will be dealt with in our panel discussion. Taking part are technical staff from seven of the leading European manufacturers of sawmill machinery for small logs. The speakers are :

Mr. Max Esterer, co-owner of Esterer Maschinenfabrik AG, Altötting, Fed. Rep. Germany, a machine designer who has introduced many technical innovations;

Dr. Hans Dietz, of Wurster & Dietz, Tübingen, Fed. Rep. Germany, specialised in designing saws and conveyors.

Mr. Jean-Claude Olsen of Ari AB, Ornsköldsvik, Sweden, whose units are
to be found all over Scandinavia and in other countries in Europe;

Mr. Cato Ebnes, of Jensen og Dahl A/S, Lillehammer, Norway, a firm
specialising in small-log conversion with circular saws;

Mr. Peter Klement, of Danckaert International, Brussels, Belgium, which
have produced units specially for small-log conversion;

Mr. Alfred Reuter, of Gebrüder Linck, Oberkirch, Fed. Rep. Germany
developer of the German profile chipper;

Mr. Heinrich Morgenstern Jr., of Kockums Industri, Söderhamm, Sweden,
which has produced a very large range of conversion units;

To open the panel discussion we have Senior Forest Officer Wolfgang Steuer
of the Bavarian State Forest Administration, Munich. He will give us information
on the supply of small-diameter logs.

STEUER: A stand of spruce planted around the turn of the century is nearing rea-
diness for felling today. A similar plantation of pine has just passed the middle
of its life and is, so to speak, in its best years. When this plantation is felled
in the year 1990 or 2030, then something will be done that was decided upon about
1900. The special nature of timber production means that the quantity, the kinds
of wood, the quality and the proportion of different types must be decided upon
a century or more in advance. Short-term alterations of production are, in practice, impossible. A further peculiarity of timber production is that the final
product is, at the same time, raw material.

The business of forestry, with its specialised and long term production,
is confronted by an industry with high productivity, mass production, automation,
rapid adaptation to the market and a high rate of growth. The changes brought
about by industry affect our whole world and have resulted in an unprecedented
rate of progress. In order to try to keep up with this development to some extent,
forest science has endeavoured to obtain a clear picture of its actual production
(reserves, new growth) and also to lift a corner of the veil which hides the future and thus anticipate future needs. Everything we plan and do today must hold
good for the future – a very distant future, as we have pointed out.

What is the position in the small logs sector at the present time and what
developments are in progress?

I will try to approach this question from the standpoint of Europe as a
whole and then deal with individual countries, proceeding from the large to the
small instead of vice versa, as is customary. I have taken the FAO studies as a
basis. The study of European wood consumption that appeared in 1963 and the revision published in 1969 show that the prognosis had been mainly confirmed by the
results and that, if anything, the estimate of development was too cautious. The
production of industrial timber, for instance, was higher than estimated.

At this point I must mention that for the designation "small-logs" as it
is understood in this discussion – logs of any length with a top end diameter of
10 to 20 cm, mainly coniferous – there is no precise definition. So I must take
industrial timber as a starting point in order to get to small logs.

Production of small timber in the German Federal Republic unfortunately
shows, in contrast to its increment on the stump, no increase but rather a decrease. This is because until now it has not been possible to rationalise the
felling of small log timber so as to offset heavy increases in production costs.
A similar situation doubtless exists in Austria, Switzerland, France and Belgium.

The result is that not all the small timber that is growing is harvested.
There are still considerable reserves of small timber in the forests.

An investigation made a few years ago into the cultivation of reserves of
timber in Austrian forests showed that measures for their care were not undertaken
to the necessary extent. The backlog in first thinnings were 80% and in second
thinnings, 33%. Similar figures probably apply to the alpine and sub-alpine regions of the Germany Federal Republic. In the Bavarian Alps the average reserves
exceed the quantity of timber felled in a year. In addition, the number of trees
taken out before general felling is usually quoted with reserve.

The question naturally arises: What can be done to mobilise the reserves?
What inducement is required?

The beneficial effect of thinning must not be underestimated. It results
in better growth and better maintenance of value and especially in increasing the
strength of the standing trees. Some success has been achieved by providing information, training, and assitance. But lasting success will only be obtained
through an appropriate material inducement.

Timber from thinnings will not appear in quantity on forest roads, until
the forest proprietor can reasonably be expected to go to the expense of felling
small timber with a top end diameter under 20 cm. The accent here is on "reasonable", which must apply as much to prices as to costs. The wood-producing and
wood-consuming industries should make a joint effort, not only to stem rising
costs, but also to bring about an effective reduction of costs. As suggestions,
I would like to mention selling logs by weight, and storage of debarked logs. It
is also to be hoped that the designers of wood-working machinery will turn their
attention to these problems when there is a greater demand for small timber.

In conclusion, I must refer to one more point. If sawnwood producers do
develop the use of small timber this can only be welcome to the forest proprietor,
since it increases the value of his product. But sawmillers then invade the raw
material sources of the pulp, paper, cellulose and particleboard producers, who
can fall back on wood of even smaller dimensions (branches, tops, brushwood, etc.)
to a certain extent. This is, however, limited.

From the FAO study for Europe for the years 1975 and 1980 I give the consumption, yield and deficit of industrial timber in Europe (in millions of m^3):

	1975	1980
Consumption	202	248
Yield	178	210
Deficit	24	38

The figures for the production of industrial timber deciduous and coniferous wood, given in the FAO study for Europe excluding the USSR are (in millions
of m^3):

Industrial timber:	1970: 125 m^3	1975: 148 m^3 1980: 175 m^3
Fuelwood :	1970: 85 m^3	1975: 79 m^3 1980: 70 m^3
Sawlogs and peeler logs:	1970: 139 m^3	1975: 146 m^3 1980: 154 m^3

In order to arrive at some conclusions for the yield of small logs with
the dimensions stated above, the following facts must be taken into consideration. Some 70% of industrial timber falls within the range of the specified top
end diameter. By cautious estimates, about 30% of fuelwood is suitable for this
type of processing. Finally, about 10% of sawlogs comes within these diameters.
Applying these percentages, the quantity of small logs in Europe available for
special processing can be calculated as (in millions of m^3) :

1975   from 142 m^3 1980   from 159 m^3

Of these quantities the proportion of coniferous wood (in millions of $m^{3)}$
is :

1975   99 m^3 1980   120 m^3

For the individual countries of special interest in this connection I have
used the data of the Timber Bulletin for Europe, which I have recalculated on the
basis explained above. I can only give the data for average fellings in recent
years. These are, for coniferous wood (in millions of m^3) :

Fed. Rep. Germany	5.5
Austria	2.5
Switzerland	1.0
France	5.1
Belgium	0.5

Production in the German Federal Republic has never followed a straight
course. Felling has increased or decreased according to the market situation, with
high or low demand. Natural disasters have also had their effect. In the year 1967,
especially, when recession and catastrophy coincided, there was a tremendous slump
in the market for industrial timber and a consequent decrease in felling. The good
market situation in the years 1970/71 resulted in an increase of felling. It
dropped to a remarkably low level in consequence of the sales crisis of 1972/73.
If the amounts felled in 1972/73 are compared with those for 1970/71 they show
a drop from 17.8 to 6.20 million m^3 for spruce alone, a reduction of 1 million
m^3 for the past two-year period. At the same time one must realise that the small
log timber is there, is being produced, since its growth rate remains approximately the same. What is not cut down remains in the wood. In the following years
a correspondingly greater quantity can be felled.

Framesaws for small-log conversion

ESTERER: Wood is a highly valuable raw material. Under proper forestry management it grows continually and is never exhausted.

Wood reaches its highest value in the form of sawntimber. The aim is always to produce the highest possible proportion of sawntimber from a given volume of round timber and, at the same time, to maximise revenue from waste products commercially as far as possible.

Sawntimber can be produced very economically with framesaws. The specific advantages of the framesaw are:

its universal application for either round or square timber

the large number of simultaneously operating saw blades

low tolerances in dimensions of sawntimber

easy handling of saw blades, their low cost and versatily with varying wood characteristics (e.g. frozen wood)

As log diameter reduces, it becomes more difficult to keep product value and production costs in an economic relationship. A framesaw cannot be merely reduced in size to handle small logs because design criteria in the saw's components are different. This applies in particular to the relationship between the framesaw opening, its stroke (s), and its speed in revolutions per minute (n). Feed rate is directly proportional (other conditions being equal) to the average saw speed (c).

$$c = \frac{s \times n}{30}$$

Mean saw speed thus increases in a linear relationship with stroke and revolutions per minute. Since the framesaw is an oscillating machine, average saw speed is governed linearly by the stroke but quadratically by the number of revolutions, that is to say, by the very rapidly increasing forces of acceleration. With present technology, even small-log framesaws can reach speeds up to 7 to 7.5 m/s and, depending on the required cutting accuracy and the kerf, a feed rate of 5 to 20 m/min. High production rates are only achieved when logs are fed end-to-end to give a high ratio of sawing hours to working hours in each shift. Time losses are often underestimated and only become apparent if a count is kept on the number of logs sawn per shift.

Production is normally quoted in units of volume of roundwood sawn or sawnwood produced. But these figures are only relative, since feed rate achieved varies linearly with the diameter of timber while the volume is a quadratic function of volume. More meaningful are figures for the aggregate length of logs passing through the saw in unit time.

High production rates are especially difficult to achieve with softwood, since small logs are usually very short and require more handling than larger timber. Standard carriages are too slow for this and roundwood has to be fed directly into the machine lengthwise by conveyors. Up to eight rollers must be used and usually fitted with grooves or spikes to maintain alignment.

A balance must be achieved between production rates and sawing kerf losses. Saw gauge thickness must be adequate, however, since rigidity increases in proportion to the third power of its thickness. Fine pitches can be used so that a

large number of teeth can be applied to the small cutting areas. This reduces
saw loading and gives a cut of better quality.

It is thin boards that are most often cut from small logs. Sawblade mountings determine the lower limit for width of cut and this is normally about 15 mm.
At this thickness boards are sufficiently strong to withstand sawing forces without breaking too frequently. For boards of 12-mm thickness special mountings are
required. Kerf loss will also be increased and feed speeds have to be reduced.
This thickness is normally obtained more economically by cutting to twice the
thickness and then halving this on a band resaw.

Small logs can be sawn very economically without major problems in a
framesaw installation planned and organised along these lines.

DIETZ: Current knowledge on raw material resources indicates that annual increment is a constant, so the increasing shortage in supply makes it a matter of
urgency for us to make the most economical use of the resource base.

This requires that the processing of tree-length timber into sawngoods
must involve cross-cutting to minimise crookedness and preparation for sawing
that takes into account diameter and quality. The sawing system chosen must satisfy as fully as possible the following demands:

1. Precise alignment of each log

2. Narrow kerf

3. Utilisation of taper in side-cut

The cost of small-diameter timber today is between 140 and 150 DMk/m^3,
delivered to the mill. The selling price for the main product is between 320 and
350 DMk/m^3. Taking a yield of between 43 and 51% as a basis, the selling price
does not even cover the cost of the raw material. In Continental European sawmills, therefore, it is essential to obtain a high yield of sidings. Sidings
currently still fetch a price of between 250 and 280 DMk/m^3. Only with this additional product can a sawmill expect to make a profit.

The price obtained for residues, in contrast to those quoted above, is
only a fraction of them, ranging from 25 to 35 DMk/m^3 for edgings and slabs.

The price paid for chips is 40 to 43 DMK/m^3 solid wood equivalent. Last of
all comes sawdust at about 8 DMk/m^3 solid wood equivalent.

From these figures it is easy to see that a Continental European sawmill
can only show a profit if recovery rate is raised by production of boards as well
as squared timber. An increase in the yield of main product or sidings of 1% will
increase profits by 1 to 1.5%. I think we cannot be reminded too often of this
fact in today's discussion.

Bandsaws have the lowest cutting loss, with about 2.4 mm per kerf. For a
board to be 24 mm thick, the initial thickness of the material must be 26.4 mm.
I will refer to the production of boards 24 mm in thickness to allow comparison.

Next lowest in kerf widths is the framesaw, with 3.2 mm. Thus the yield in production of 24 mm-thick boards drops by 1.5% compared to the bandsaw.

Conventional circular saws have a kerf considerably wider than that of the framesaw, so that, in comparison to the bandsaw, a loss of yield of up to 7% occurs in the production of 24 mm-thick boards. Recent developments in circular saws using multiple arbors, with saws penetrating the cut by stages and from both sides, have reduced kerf loss to about that of the framesaw. Face-cutting canters have no cutting loss in this sense, since they do not divide the material. The best of the actual cutting machines is thus the bandsaw. It must not be forgotten, however, that the cutting accuracy of a bandsaw is normally less than that of a framesaw or circular saw.

I now turn to actual yield obtainable from all logs in the diameter range 12 to 22 cm.

I will assume, here, logs with an average length of 4 m and an average taper of 1 cm/m. For volume calculation, an average diameter of 17 cm will be assumed. Calculation of yield with the main product, however, involves only top diameter, which would be 15 cm. We can consider both production of 24-mm thick boards and square timber by dividing the entire dimension range into two with top diameters of 10 - 15 cm and 16 - 20 cm.

In the smaller category, production of square timber gives a yield of about 43% irrespective of the type of machine used. In production of 24 mm-boards, yield with a narrow-kerf bandsaw is 45.5%, with a multi-blade framesaw 44%, and with a circular saw between 43 and 39%.

In the larger log category, with top diameters between 16 and 20 cm and the use of a chipper-canter, yield of square timber increases to 51%. With a bandsaw producing 24-mm boards, it increases to 63.5%. Yield with a multi-blade framesaw is 62%, and with a circular saw it is between 60% and 54%, irrespective of whether the 24 mm-boards are produced in a double or single cut.

These figures show the bandsaw as the one that stands out in cutting logs in the smaller category.

In the larger category, yields given by the bandsaw, framesaw and circular saw are better than that given by the chipper-canter. Here, use of the chipper-canter must be restricted to removal of slab and one of the other saws has to be used on sideyield to give economic results.

Using these recovery rates we have calculated the costs of investment, power, and labour, for a sawmill equipped with various combinations of machines to give an annual production of between 15,000 and 30,000 m^3. Sawing costs in DMk/m^3 for main product and sidings was calculated without taking into account revenue from sale of chips, slabs, and sawdust. The following machine-combinations were considered:

1. Multi-blade framesaw line
2. Double-cut chipper-canter
3. A four-face cutting Chip-N-Saw machine
4. Circular saw line with pre-cutting saw and resaw
5. Combined chipper-canter and log conveyor sawmill with circular
   trim saws and chipping edgers (as proposed by Bruxaholms, Sweden)
6. Quad-bandsaw and two double-trimmers

Log sorting equipment is installed only in the framesaw line. A debarking machine and log infeed system is included in all the installations. Sorting equipment for side yield will be provided if sidings are produced, together with stacking facilities for the main product and sidings.

In the 10-15 cm top diameters category, the chipper-canter produces best results. It is followed by the framesaw and the Chip-N-Saw installations, with about 10% higher sawing costs. Sawing costs with the circular saw and the combined quad-bandsaw with the double trimmer installation are approximately 20% higher. Sawing costs with the combination chipper-canter, multi-blade bandsaw and double trimmer are between 35 and 40% higher than the chipper-canter despite high recovery rate. In this diameter range, a chipper-canter should be used.

In the larger diameter category 16-20 cm top diameters, the framesaw line gives costs that are 2% better than the Chip-N-Saw. The chipper-canter, circular saw, and combination quad-bandsaw with double trimmers all give a result that is 10% below that of the framesaw. Once again, the worst result - 25% higher sawing costs - is given by the combination of canter, bandsaw, and double trimmers, despite the fact that again it provides the highest recovery rate. In both instances high investments prevent this installation from being more profitable. On economics, therefore, only the multi-blade framesaw can be considered for cutting timber with a top diameter of 16 to 20 mm and for all diameters larger, up to about 60 cm.

In conclusion, the conventional framesaw still remains the optimum solution in Continental Europe for sawing timber with a top diameter greater than 16 cm. This conclusion is, naturally, based on a mill with a relatively small capacity, but the range 15,000 to 30,000 m^3/yr capacity is accepted as optimum for Continental Europe. If there were a greater area of forest here, as in Northern Europe, or in the USA and Canada, it is quite possible that other methods of sawing timber could be used economically.

Processing of small logs with circular saws

OLSEN: The basic assumption behind the Ari-System is that you want to gain a maximum yield of sawn lumber out of each log that is transported to the sawmill. This is as a rule what the sawmiller demands from his equipment.

Furthermore, when sawing small logs you have the additional requirement of high capacity in terms of number of logs processed per unit time.

The modern circular sawmills with multi-blade machines and resaw can meet these requirements to a high degree. The recovery of sawn wood will be high thanks to individual log treatment and resawing of the side yield in a separate resaw. The selective sawing method allows sawing of precisely that dimension which is most sought after with regard to the quality of the wood.

Increased attention is now being given to lumber processing (sawing and then sorting) to grade because of the improved value of manufacture this gives. In the modern circular sawmill - not least by means of the resaw - the full potential profit implicit in the idea of sawing to grade can be achieved.

An occasional misjudgement, an incorrect alignment or dimension setting
at the head saw can be compensated to a great extent in the resaw. Every experienced sawmiller knows how important this advantage can be.

The modern circular sawmill comes in different sizes designed to process
up to about 350 logs/hr depending on log dimension, dimension of sawn lumber and
so on. The main features of circular sawmills of the Ari-System are:

1. High recovery of sawn lumber due to the selective sawing method;
2. Low dimension tolerances, and straight cutting, even when sawing
   frozen logs;
3. High feed speeds;
4. Even and smooth cut surfaces;
5. Simplicity in design and lightweight machines, keeping machine
   and foundation investment down;
6. Mill layout readily modified to suit new requirements.

The system allows good adaptability to the conditions of the individual
sawmill with regard, for example, to log dimension, log quality and dimension of
the sawnwood. In a small-log sawmill of this kind also larger logs up to a diameter of 50 cm can be sawn without loss of recovery.

This flexibility of the circular sawmill can be of great importance equally to the small-log sawmill as to mills in regions where the log supply can vary
considerably in average log size from one year to another due to variations in the
demand of pulpwood.

Small-log sawing with circular saws

EBNES: Let us start immediately with an example of a customer's demands:

Required production per year:
35 000 m^3 round logs (input);
One shift operation, 220 working days;
Average log: 160 dm^3;
Average length of log: 4.5 m;
Indicated top diameter: about 18 cm;
Production to increase up to 50 000 m^3 input/yr.

Sawing on the curve for crooked logs is important for this customer.

We constructed this mill and the flowline is as follows:

The logs are pushed   the log intake conveyor. No sorting of the logs is necessary. All dimensions can be mixed - one of the advantages of a circular sawmill
with a double log slabber.

The first break-down machine is the double log slabber where two slabs are
cut simultaneously. If the logs are crooked the log should be sent through the machine with the curved side upwards. This is the first step of practising curve
sawing, which is very important for this sawmill. I shall revert to the curve sawing later. Behind the double slabber there is an automatic slab operator and slab
kick-off device. Slabs fall onto a cross conveyor for transport to a double edger.
More on slabs later.

The cant is turned 90° and pushed over to an automatic infeed and centering device in front of the split saw. This device keeps the centre line of the
cant in the line of the circular saw of the split saw. This splitter always cuts
the cants along the pith. Even curved cants can be sawn along the pith instead of
diagonally. I shall revert to this later.

The two half-cants are brought to two resaws, one mounted behind the other.
One operator pre-selects the dimensions for both machines on a console. He can resaw with just one machine or with both. The planks go out for sorting, smaller
slabs go directly to the chipper, bigger slabs to the double edger together with
the slabs from the double log slabber. The resaws are strong enough to follow the
curved half cants.

Practising curve sawing on the resaws we follow the fibres of the half-
cants. We end the curve sawing here and have saved timber.

The operator of the double edger pre-selects the dimensions on the double
edger as well as a slab resaw behind the edger. Edged slabs are automatically fed
into the slab resaw which the single operator controls.

If squares are wanted out of some logs the first two slabs are cut as
usual on the double log slabber. Instead of sending the cant to the split saw the
operator can let the cant fall down on to the cross conveyor together with the
slabs. The operator of the double edger sends the cant through the edger and thus
makes squares, which, of course, do not go through the slab resaws.

The operator of the double edger has a cross-cut saw at the infeed bench
to trim off the thin ends of the slabs. On the cross conveyor for slabs going
from the double log slabber to the double edger there is also a cross-cut saw. If
the mill manager wants to speed up for a certain time, and, for instance, he has
a large order for bigger dimensions, he can have an extra operator just for cross-
cutting.

In this sawmill there are six machines and three operators: one for the·
double log slabber, one for the two first resaws and one for the double edger and
the slab resaw.

By practising curve sawing you follow the pith of the cants and do not saw
diagonally. You cut along the fibres which means better planing results later on.
You save timber and, according to several sawmill owners in Norway, they save a
lot of money each year by practising curve sawing.

Bandmills for small-log processing

KLEMENT: I propose to deal with ways to cut squared logs economically by bandmill
headsaws and resaws.

Discussion is based on the following production data and dimensions:

    Annual one-shift capacity: 3,000 to 15,000 m^3
    Average diameter: 12 to 22 cm
    Log lengths: 3 to 6 m
    Type of wood: Coniferous

Factors involved here are many, and include daily rates for roundwood-processing, log lengths, and total metres of saw-cuts. Further, these data vary with the sawing programme. For this reason I will break down the total capacity range to be discussed into three.

Annual capacity from 3,000 to 6,000 m^3

For cutting squared logs in this capacity range we recommend a modern design of pneumatically-controlled headrig followed by a double circular edger.

Initial breakdown of round logs is done by the headsaw. Depending on diameter and sawing programme required, it either produces a square cant or a flitch which passes to the circle edgers.

The band headsaw is operated by one man with control of a transverse conveyor, a hydraulic or pneumatic loading and turning mechanism and carriage.

The double edger requires a depth of cut of at least 160 mm and be able to produce boards and cut the flitch without difficulty.

The edger can effectively be combined with a retracting trimsaw mounted under the feed table to allow trimming of waney ends before edging.

This installation would also require conveyor systems for centreyield from bandsaw and edger to take it to one greenchain and a second system to take boards to a second greenchain.

Annual capacity from 6,000 to 10,000 m^3

Possible installations for this production range depend on cutting programme. They are :

1. Band headsaw - band resaw - twin circle saw edger.
2. Band headsaw - large-diameter circular saw - double circle saw edger.
3. Large-diameter circular saw - band resaw - double circle saw edger.

Again, cutting programmes determine the final choice of sawing units.

In line 1., initial breakdown is always done by the headsaw. Flowline should allow workpieces from the headsaw to pass to either the resaw or edger or directly to the greenchain.

Production line 2. is particularly suitable for small logs. Flowlines must again allow all likely sequences.

Installation 3. has all initial breakdown and part of subsequent sawing carried out by a large-diameter circular saw. The resaw handles slab and centreyield from the large-diameter circular saw.

Boards from the resaw pass to the edger, which can also handle production from the large-diameter circular saw.

Annual capacity from 10,000 to 15,000 m^3

Here a modern band headsaw is used in combination with a high-yield resaw and double circular saw edger. In this installation the resaw becomes the production machine. It is equipped with an extra infeed for cants and a linebar so the saw can cut both cants and slab.

From the resaw boards can be fed to the edger. Squared timber from the headsaw and edger can pass directly to the greenchain.

Slab is separated downstream at the edger requiring only boards to be handled on the greenchain.

Machine specifications

Band headsaw

Diameter:	1,400 – 1,600 mm
Saw steel width:	200 mm and up
Carriage length:	6,500 mm and up (with 4 knees)
Saw opening:	600 x 600 mm
Carriage speed:	100 – 150 m/min (infinitely variable)

Specifications for the resaw are similar to those for the band headsaw.

Band headsaws have improved viability not only in equipment within the mill but often in the logyard, since unsorted logs can be cut. The advantages of bandsaw installations for a planned sawing programme can be summarised as follows:

1. High-quality finished product
2. Flexibility in cutting programme
3. High recovery rates

Interest in recovery rates is growing as raw material shortages become more apparent.

Profile chippers in small-log conversion

REUTER: A statement on output and economic efficiency in the sawing of small logs can only be made if the most important data for each particular case are available. It would go beyond the scope of this discussion to consider output and economic efficiency of the various sawing techniques for all possible lengths, diameters and sawing programme as well as use of sideyield.

The following are the most important data for the determination of each individual case:

1. Required output
2. Cutting programme and the number of cuts
3. Log length
4. Log diameter
5. Type of wood and straightness of the stems
6. Production of side products

In general, the following machines are used for cutting small logs:

Framesaw: The most universal machine for the economic cutting of small logs is the multi-blade framesaw. Unfortunately the sawing speed and therefore the output of this machine is limited, since with small logs the number of saw-cuts is small.

Bandsaw: With a small number of cutting kerfs, two or four per stem, a double or quadruple bandsaw is interesting, provided that a good centering and feed device sees to alignment, slab separation and other processing.

Circular saw: A circular saw is of limited use in cutting small logs, since the cutting depth is limited and the cutting kerf is too wide.

Chipper-canter: Logs can be easily and rapidly cut to geometrical profiles with this machine. To obtain the full benefit from the chipper-canter, it is important to know that cut cants greatly facilitate subsequent cutting with multi-blade saws.

The possibilities of the various methods of sawing are best and most quickly shown by a few examples, arranged in order of the quantity of timber cut annually:

Example 1:

>Data: 5,000 m^3/yr, round timber 2 - 6 m long, diameter of top end 10 - 15 cm, sawn into squared timber.

>Recommendation: Pre-cutting with a building-timber circular saw, final cutting possibly with a double edger.

Example 2:

>Data: 10,000 m^3/yr, round timber 3 - 8 m long, diameter of top end 15 - 20 cm; sawn into boards.

>Recommendation: Pre-cutting with framesaw, final cut partly with framesaw, partly with double edger.

Example 3:

>Data: 20,000 m^3/yr, round timber 2 - 6 m long, top end diameter 10 - 15 cm, sawn into squared or framing timber.

>Recommendation: Pre-cutting and final cut with a chipper-canter, separation with circular saw or bandsaw.

Example 4:

>Data: 30,000 m^3/yr, round timber 1.25 - 2 m long, top end diameter 10 - 25 cm; sawn into pallet boards.

>Recommendation: Pre-cutting with a chipper-canter, subsequent cuts with another canter, followed by a separating circular saw and a double-splindled multi-blade circular saw with two feed lines.

Example 5:

>Data: 60,000 - 100,000 m^3/yr, round timber 3 - 6 m long, top end diameter 15 - 22 cm; cut into planks and boards.

>Recommendation: Pre-cutting with chipper-canter followed by a double bandsaw, final cut with another canter and a double or quadruple bandsaw, or final cut with two framesaws with automatic feed and appropriate machines for processing the sidings.

Combination saw systems for small logs

MORGENSTERN: I would like to deal first with the various sawing techniques and then with the possibilities of their application.

Technical developments in the sawmill machine industry have gone ahead by leaps and bounds and the subject of this discussion will no doubt have been chosen with the idea of receiving some indication or other as to whether it is really true that the framesaw will be supplanted by the bandsaw or modern circular saws, or whether even the cutting machine will be superseded by circular saws.

A glance at the world's largest producers of sawntimber shows that all the different sawing techniques – framesaw, bandsaw, circle saw, canter – compete more or less peaceably with each other, the justification for the employment of one or the other being provided by specific local conditions. At Kockum, which formerly concentrated on the manufacture of framesaws and canters, we have complemented our production programme in the course of the past year by the inclusion of circular saws and bandsaws.

With regard to circular saws, it is our opinion that their limitations are only too obvious, due to their wide kerf and their inadequacy for use in fully automatic production lines with a feed rate of 50 m/min and continuous in-feed. The firm of Adco West in the United States, which has belonged to the Kockum n group for some time, has produced a solution of this problem with the introduction of extremely thin, water-cooled saw blades which set a new standard of efficiency.

In respect of framesaws, we at Söderhamm are in the course of modernising these machines as well. Feed rate has always been the bottleneck of the conventional framesaw, and we are replacing these with a framesaw with smaller kerfs and higher feed rates.

When it comes to bandsaws, Kockums can also supply a modern machine. The "high-strain" bandsaws manufactured by Letson and Burpee in Canada produce a narrow kerf by operating the blade under increased tension. The combination of high saw tension and a thin saw guarantees greater cutting accuracy than is possible with the conventional system.

The canter (figure 1, see view 1 page (i) has, with 150 machines in use in Europe, achieved a leading position in the sawmill industry, not only as a machine for small logs but also as a reducing machine (face-cutting reducer).

The question of the economic use of one or the other sawing machine is always determined by local conditions. The situation in the Continental European sawmill industry can, perhaps, be summarised by saying that it suffers from a chronic scarcity of raw material (especially as regards small logs) on the one hand and that, on the other hand, the sawing programme is extremely varied.

These circumstances have led to the development of a sawmill designed around a relatively powerful headsaw. Up till now these machines have operated with a low co-efficient of technical utilisation.

What measures can be taken to rationalise the Continental European sawmill?

One obvious improvement is to relieve the pressure on the existing main machine by the installation of a separate production line for small logs (see figure 2 next page). The direct effect of this measure is a better utilisation of the headsaw, but it is accompanied by the danger that the sawmill must constant-

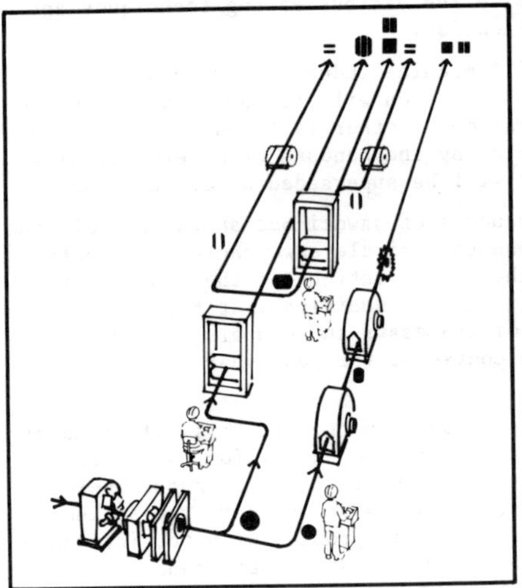

Fig. 2 - Addition of small log line
to relieve pressure on main machine

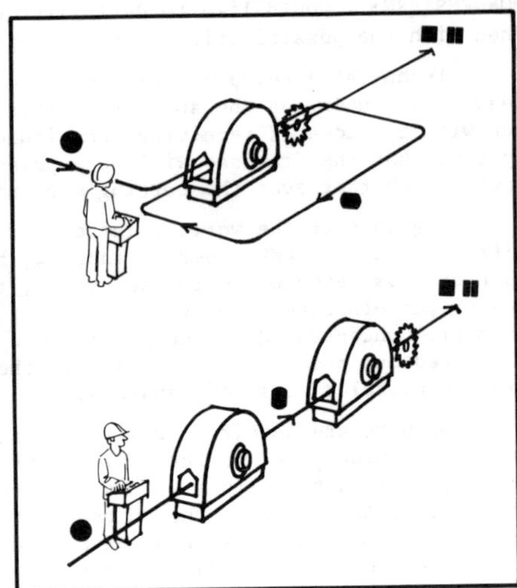

Fig. 3 - Canter for small logs
processing

Fig. 4 - Variation with two machines
in line

Fig. 5 - Combination of canter and
framesaw

Fig. 6 - Possible layout with
installation of a canter

Fig. 7 - Reducer bandmill

ly feed and fully utilise two production lines.

In our view, better rationalisation can be effected by the use of the face-cutting reducer, especially in Continental Europe. The idea is to prepare and shape the logs with the canters in such a way that sawing in the main machine will be simpler and quicker, that is to say, more efficient. Other advantages are a smaller labour force, the option of automatic infeed – to framesaws, for instance – direct yield of the by-product in the form of chips, and a greater flexibility in the operation of the sawmill, which is such an important question in Continental Europe.

Figure 3 (see opposite page) shows a canter in its simplest application in a production line for processing small logs.

Figure 4 (see opposite page) shows two of these machines in line. This basic form of installation will recur in the layouts to follow.

Figure 5 (see opposite page) shows a combination of canters and, as an example, framesaws, in which the canter is used both as a production machine for processing small logs and also as a reducing machine, that is, a pre-shaper.

Rationalization of these types does not necessarily require construction of a new sawmill. Figure 6 (see previous page) illustrates possible layout of a conventional sawmill after installing a canter.

The combination of cantering side cutters and bandsaws has resulted in the production of the reducer bandmill shown in figure 7 (see previous page). A unit with two side cutters and four bandsaws can make six cuts in one operation. In other words, the reducer bandmill can give as many cuts as the framesaw.

Compared to the conventional framesaw, the reducer bandmill gives a higher recovery with its milling cutters (no production of sawdust) and narrow kerf bandsaws, and the machine has been very well received. The improved utilisation possible with a reducer bandmill makes particularly good sense for Continental European sawmills with raw material so scarce. Further, with feed rates adjustable up to 50 m/min, a single line can handle medium-sized logs as well as small sizes.

Economic sawing of small logs with a single headsaw cannot be achieved, be it a framesaw, bandmill, chipper-canter, or circular saw. The answer, as I see it, (see figure 8) lies in a combination of the chipper-canter with one of the other three.

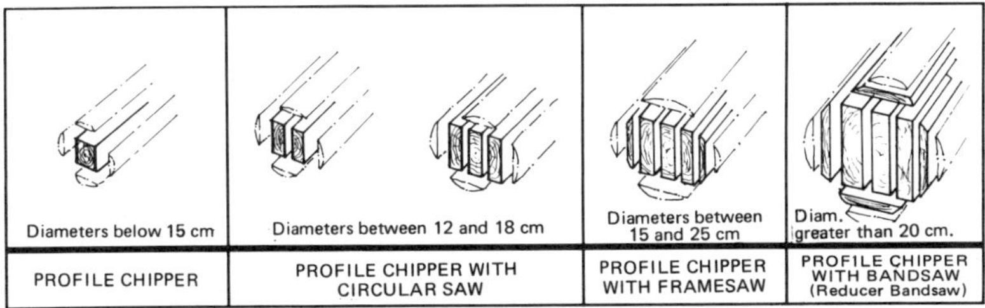

Fig. 8 - Chippers for removal of slab give the optimum sawing systems when used in combinations as shown, except for the smallest log sizes, where chippers should be used alone.

FRONIUS : Thank you for your presentations. They have contributed new ideas and data. However, mill capacity utilization rates and mean log diameter averaged over the year are vital for every installation, and I would like to add some further data on that. Figures are taken from a study I made.

Maximum outputs for the three types of saw line, assuming a 17.5 cm middle-log diameter and a 1,850-hr working year are given below, with the range of hourly outputs for the mid-log diameter range 12 to 22 cm. Sorted logs for the framesaw and unsorted logs for the circle saw and chipper-canter are assumed.

	diameter	feed speed	utilization rate	output/yr (approx.)	output/hr (approx.)
Framesaw sorted logs	17.5 cm 12-22 cm	12 m/min	0.8	12,500 m^3	(6.93) 3.26-10.94 m^3
Circle saw unsorted logs	17.5 cm 12-22 cm	35 m/min	0.5	23,000 m^3	(10.11) 4.75-15.96 m^3
Chipper-canter unsorted logs	17.5 cm 12-22 cm	45 m/min	0.5	29,000 m^3	(12.99) 6.10-20.52 m^3

Unit costs for sawing operation only

	annual output	cost per m^3
Framesaw	5,000 m^3 12,500 m^3	54 DMk. 38.15 DMk.
Circle saw	5,000 m^3 23,000 m^3	43.66 DMk. 18.46 DMk.
Chipper-canter	5,000 m^3 29,000 m^3	36.48 DMk. 13.31 DMk.

To complete this picture one should mention the recovery rate by main products, side boards, sawdust and chips. All these products have a market and also a market price. A few figures here

	main product	side boards 20 mm	total recovery	sawdust/3.8 mm kerf	Chips
Cants 8 x 8 cm with squared edges	40.3	12.5	52.8	19.4	27.8 %
Cants 14 x 14 cm with squared edges	50	20	70	11.7	18.3 %
Waney boards 24 mm x 10 cm	34.8	25	59.8	17.9	22.3 %
Waney 14 cm boards	43	22.2	65.2	17.5	17.3 %

It is clear from the above detailed figures (for 12- to only 20-cm mid-
diameter logs) that the best type of machine or combination in each individual
case can only be determined by individual evaluation of each plant taking into
account all the influencing factors.

To sum up I would like to point out that the profile chipper is preferable
for the smaller average diameters of log between 10 and 15 cm and particularly
when wane is acceptable in cant cross-section and production of boards from sideyield is minimised.

For larger diameters, and especially when products with a rectangular cross-
section are to be produced, large sideboards are to be reckoned with. At today's
market prices the amount of sideboards produced is often decisive for the profitability and even the viability of a sawmill. Thus it is preferable in each case
to use a sawing system with a separate chipper.

It is important that capital investment costs are in the correct proportion
to production to ensure economic viability. In all cases, economic viability and
profitability are always more important than modern equipment and technical per-
fection.

I hope this discussion between specialists has given you some valuable hints
for your own daily problems and that it gives the machine manufacturers some new
ideas for their own research work.

Discussion

BRAUN: The question of diameters has received much attention and I thoroughly
agree with Herr Dietz in distinguishing sub-categories within the diameter range
we are discussing. I found I disagreed with Herr Fronius regarding chipper-canters
with his figure of 46-55% recovery rate. More realistic in my experience is the
range from 42 to 45%, and this with top diameters only up to 18 cm.

FRONIUS: In fact, I can confirm the figures you quote, when I refer specifically
to the diameter range of 18 cm. This gives me a recovery rate of between 40 and
46%. It is evident from this that information on diameter allows very precise comparisons.

TUNKR, L., Kralovopolska Strojirna, CSSR: The question of realizing maximum or
optimum yield with small logs is a specific one in our sawmills in Czechoslova-
kia. We operate, for the most part, in integrated mills where sawdust can be
used for production of board or pulp. The question that arises concerns the relative merits of mechanical and chemical processing and how to find the border
lines between them. In the case of cutting boards it is apparent that conversion
costs are about DMk. 140 and the selling price about DMk. 370. Unit costs for production of core stock are DMk. 170 with a selling price of DMk. 450.

6

Sawmilling and log input practices in Canada and the USA

By James C. Wallace, Vice-President, Miller Freeman Publications, San Francisco, USA, and Publisher, WORLD WOOD and FOREST INDUSTRIES magazines.

The following is the text of Mr. Wallace's commentary
that accompanied some 100 colour slides, certain of which
are reproduced at the back of this book, starting on page i.

Our plan today is to acquaint you with modern sawmill practices in North America. Our picture tour will take us into British Columbia, Canada, and the western portion of the United States including the states of Washington, Oregon, California, and Montana. I am indebted to the staff of FOREST INDUSTRIES and WORLD WOOD for this material.

For the tour we will "zoom" in on the far western (see page i, view 2) section of the United States and Canada. We call this small part of the world the Pacific Northwest.

By way of explanation, I am not a forester nor a sawmill expert; but, sitting with me is Carl Mason of H.C. Mason & Association, Portland, Oregon, an experienced sawmill man. He will answer your technical questions.

Our discussion today will encompass, first, the natural resources in North America; second, log harvest. Each of these serves as a background to the major topic - that of lumber manufacturing techniques. We will also touch very briefly on the market for lumber products, because the market has a large impact on manufacturing.

Referring back to the map - softwood forests of Western North America combine managed stands and virgin areas. However, nowhere in this area is management as intensive as is common in Western Europe.

North America has about one third of the world's softwood forest and harvests about 40% of the world's softwood. About half of this amount - and a high percentage represents high quality wood - is in the far west part of the United States and Canada. Although the United States exports both logs and lumber; it may surprise you to learn that this country is a net importer of lumber. Nearly all the lumber import comes from our neighbour to the North, Canada's westernmost province, British Columbia.

Let us look into the timbered areas of the United States' West. The inland areas that grow timber follow the mountains and are high elevation forests, relatively open. Pines and larch predominate, but several other genera are common. For example, the area adjacent to Pacific Ocean has heavy rainfall, moderate temperatures and dense forests. The climate is much like England. Douglas fir predominates. There are also dense stands of hemlock (tsuga), true firs (abies), spruce (pinea), cedar (thuya), and redwood (sequoia). So much for the natural resource. Now let us turn our attention for a moment to harvest.

To introduce you to the inland area of the state of Washington we show a rubber tyred skidder working in a stand of lodgepole pine (view 3). This type of small timber - we call it second growth - is also typical of the pine that grows well and rapidly in the southeastern parts of the United States and in the eastern provinces of Canada. Now we show log loader in the same general area. Underbrush, you will note, is light in this region.

The camera now moves to the far west coast region. This is a Douglas fir area, very close to the Pacific Ocean.

An overview of a Douglas fir forest. This species must be clear-cut to regenerate properly. In the center of the slide is Mt. Shasta - located in northern California. This mountain is more than 4,200 meters high and is part of the Cascade range, the coast mountains extending North to South that prevent most of the rainfall from reaching the inland areas.

Now we show a stand of Ponderosa pine located in the central section of Oregon. Heavy equipment is typical of this area. Several logging methods are used. This is a track-type skidder with bulldozer blade, typical of the tractor used in yarding logs, and sometimes involved in road construction.

This view shows a modern steel tower used in western logging, which replaced the old-fashioned spar tree. This tower is just over 33 meters high. It is used in this fixed position for yarding logs from the incline beyond. The unit is mobile, and can be moved to a new location after the rigging is taken down and the tower telescoped and tilted back on the body of the yarder.

A choker-setting crew is hooking logs onto the line here. The logs are then yarded to the landing. At the landing the logs are decked - or piled - for loading on trucks and trailers. This is a typical loading operation - loading the logs on trucks and trailers for transport to nearby mills.

And now, a sawmill yard, where the trucks and trailers are unloaded and wait for their return trip to the logging area. Note the truck with three trailers. Hauling with more than one trailer is permitted only on private roads, that is, roads owned, build, and maintained by the mills.

Now we show an unloading process. The unloading is generally done by a "log-stacker". The function of the stacker is to sort logs at the millyard by species, size and use. For example, the larger logs - and those with few knots - go into the manufacture of plywood. Here we see a rotary cutting plywood lathe. A moment ago, I mentioned Douglas fir, hemlock, larch and true firs. These species generally are destined for structural uses. The pines generally go into panelling, sash and doors, moulding, shelving, etc.

In British Columbia, about 95% of the timber is owned by the Provincial government. In the western United States about half of the timberland and 60% of the inventory is owned by the federal government. The rest - including the best growing lands - is privately owned.

The most productive timberland area in the West, outside British Columbia, is in the states of Oregon, Washington, and California, where, as I said, the federal government is the principle landowner. This contrasts with the South which is the pine growing area, where most of the timberland is in private ownership. In the next 25 years, the South will provide more and more of the lumber and fibre raw material for the USA.

The sawmill industry of North America.

Now the main part of our theme - the manufacture of lumber.

Many North American mills are quite large. Notice the size of the log deck behind the mill in this view which is in the southern interior of British Columbia (see view 4).

Large decks must be maintained in many western areas as weather prevents access to logging areas for part of the year. The very largest North American mills are capable of cutting up to 472,000 m^3/yr. Most, however, would fall between 35,000 m^3 and 118,000 m^3 yearly. Sawmill is at the rear on the right of the view (again, 4).

Lumber, sorted at the greenchain (to the left of the mill in view 4), goes to the dry kilns at far left; then through the planer in foreground to the rail siding.

Some North American mills continue to store roundwood in log ponds and where a natural lake is available it is often used for log storage.

Logs may be barked and bucked to desired length, then returned to water storage. However, despite water availability, most logs are stored on land or "dry-decked" as perhaps 12 times as much wood can be stored in the same surface area.

After logs have been barked and sorted to size and grade, they are decked. Very large rubber-tired stackers unload trucks, sort, stack, and feed the mill, working with the conveyor.

This view (5) emphasized the large capacity of the stacker. It works here at gathering logs which have arrived by log raft from the upper reaches of the Columbia River.

Bucking or cross-cutting arrangements for North American mills are various. Guilloting cut-off saws are one system.

Use of scanner-computer combinations to determine log dimensions and dic-

tate cutting solutions, are rapidly becoming common in medium-sized U.S. mills.
They are used principally for small logs, however. Here a small hemlock log
moves from the ring debarker through the scanner.

Scanner rely on reflective light to measure log length and diameter. Optimum sawing programme for each size log is already stored in the computer which
automatically actuates the setworks. The sawyer can override the computer's decision manually, however. This view (6)shows in a general way the interior of
large U.S. mills. Note that almost all work-in-progress is transported by rollcase or transfer chain. Handcarts are unknown and lift trucks are not used on
the mill floor.

Headsaws for large and middle-sized logs (view 7) in western North America
are mostly bandsaws. The circular saw mounted on a carriage and the framesaw
(or gang saw, as we call it) are obselete as headsaws. This is a 3-meter double-
cut bandmill. The man in the center is the offbearer. The sawyer is out of
view in the enclosed cab at the right. A closer view, this time showing a Douglas
fir slab as it comes off the log. These logs are old-growth Douglas fir, often
350 years old and frequently 600 years or older. They are more likely to contain
heart rot or other defects than young logs. But they also contain near the outside, far higher quality clear wood, free from knots. The sawyer must be able
to turn the log several times as it is opened; and its defects - or high quality
areas - become apparent to him. A high-strain bandmill capable of withstanding
stresses up to about 15,000 kilograms. Some of these high strain bandmills operate with saw thicknesses as low as 2.4 mm. They are not typical however. Saw
thicknesses of 3.1 to 3.8 mm would be more usual for conventional strain mills.

A popular approach (see view 8) to initial breakdown for small logs in North
America is the chipping headrig ("Beaver or Chip-n-Saw"), which chips the outside slab surfaces instead of sawing them off, but nevertheless, creates a lumber profile for the cant. The larger of these machines also make any interior
saw lines needed in the cant in the same movement. Thus the outside sawkerf,
which would have been lost to sawdust, is saved in chips.

This view (9) of a Chip-n-Saw shows the outfeed, with a 10 x 20 cm hemlock
cant.

Another approach to reducing sawdust and increasing chips in the area formerly lost as saw kerf, is the headrig chipper or headrig slabber, used with
a bandmill. Here we have a Passavant slabber ready for a cant which is being
turned. This is a six-knife machine with a 61 cm cutting circle. Reportedly,
it increased production for this California mill by almost 17%.

Here view 10)a slabber is chipping at the same time the bandsaw is working
on the log. Swelled butts are taken off at the headrig and the resulting whole
log chips are reported of higher quality. Elimination of one to three lines
per log increases the bandsaw's capacity. Note the automatic offbearer slat-
bed in foreground. This approach sometimes makes it possible to eliminate the
tail sawyer.

As better machinery increases the opportunity of mills to improve recovery
and grade, the effort to create an environment for the skilled mill worker to

perform to his best ability also is improving rapidly. Machine operator cabs, such as this (view 11) are heated and cooled, exclude dust and noise, have tinted safety glass tilted for the best view, and they exclude glare. Many include closed-circuit television monitors to permit the sawyer to see down the track, assuring the safety of his offbearer. Padded seats tilt and adjust up-and-down and back-and-forth. Cabs are also being installed in some mills for edger,resaw and trimmer operators, as well as for head sawyers.

Most commonly, American edgers are of the double arbor type. Here we show (view 12) a typical Schurman brand and its outfeed. This is the outfeed (view 13) of a 152-cm wide gang edger using 55-cm saws with 3.1-mm kerf.

Many types of resaws are becoming popular in the Western region, but the twin band is gaining favour rapidly (view 14). It can accept any piece on which one face has been sawn - yet it uses thinner saws than the headrig requires.

Many mills now cut with double and triple band resaws to gain the resaw's kerf savings. Most resaws can be fed directly from the headrig, and from the edger, and also have merry-go-rounds to hold off material when the saw is "burried".

North American trim saws are generally multiple circular saws spaced at 61-cm increments. This trimmer (view 15) will accept 8-meter lengths. Pieces are fed through singly and an electronic memory system drops the correct saw when the piece arrives.

Lumber, to be shipped green, or which has come from the kiln, goes to the planning machine . All American lumber is dressed on all four sides.

Much manual labour has been saved in U.S. mills by large lumber sorters with automatic memory systems. In this view (16) we see a 14-tray system. Sorts can be made to width, to length and to grade. Tray sorters have unloading tipples. Bins are often unloaded by fork lift truck. The more traditional method is on a green chain where sorting is by hand. Fork lifts are the common stacking and loading vehicles in U.S. mill yards.

Most strapping is automatic. Here (view 17) is a Moore-brand automatic lumber package strapping machine.

Most lumber is shipped in rail cars. Sometimes lumber is shipped across the U.S. in open flat cars and, therefore, is wrapped with heavy, weather-proof paper. The U.S. is over 5 500 km from West to East, but lumber travelling from northern British Columbia to Florida, the southeast tip of the U.S.,would have to go perhaps 8,000 km if shipped in a straight line.

I wish to digress at this point, getting away from the sawmill and the lumber processing techniques to discuss the market.

The big market for wood in the United States and Canada - the market which is going to put a heavy strain on our forests of tomorrow - is that represented by two strong influences: housing and fibre for pulp and paper.

Both single-family and multi-unit dwellings in North America are constructed basically of wood. Frames are in lumber and plywood sheathing is used for wall covering.

What is the housing market ? Here is a chart (view 18) dramatizing by

statistics the number of housing starts that stimulates or slows down the manufacture of lumber in the USA.   In 1973 our housing starts totaled 2.2 million
units.   This gave the lumber industry a great stimulation last year, some of
our lumber and plywood companies experienced their most profitable years.   In
1974 experts anticipate that housing starts will decline to approximately 1.8
million.   This has slowed the nation's lumber industry economy.   There are other
influencing factors: namely, the rising costs of mortgage money; the rising cost
of land, and the rising costs of harvesting and manufacturing wood products.

Another vital market impact on the manufacture of wood products is that
of fibre, the raw material for pulp and paper, particleboard and hardboard.
There will be growing competition within the industry for the logs, logs destined for fibre, logs destined for plywood, logs destined for lumber.   At most
mills the decision will be made on the basis of the most profit from the log,
as that factor is related to the type of timber available to that particular mill.
In North America we now have pulp mills going into the sawmill business simply
because some of their logs will yield more profit as stuctural studs than fibre.
A Canadian forestry official recently said that the production of fibre in Canada
would triple between the present day and the year 2000.   In order to meet this
tremendous demand, North America will see a rapid expansion of what is now being
called whole-tree chipping.

Here we see a Morbark Chipharvester (view 19).   There are reportedly 90 of
these machines now operating in the woods of the USA and Canada.   The Chipharvester handles the whole tree - main stem, limb and foliage.

Wood has become much more valuable in North America at the same time that
public pressure for clean air has grown.

During the past few minutes you have had a very rapid tour of the sawmill
industry in the northwestern part of North America.   Should you desire to visit
this region in the future, you should plan to start from such cities as Vancouver, B.C.; Seattle, Washington; Portland, Oregon; or San Francisco, California.
Any of these cities would provide a good base and starting point.   Most of the
mills we have visited are accessible by plane and all are easily accessible by
automobile.   Such a trip would take you a week to ten days.   Thank you very much
for this opportunity to speak to you.   We are now ready for your questions.

Discussion

THUNELL (Prof. Bertil, Swedish Forest Products Research Laboratory, Stockholm,
Sweden): I have four questions to ask :

1. To what extent do you think you will use sanding instead of planing, when
   dressing the lumber?
2. To what extent do you think that you will break off the corners of the
   lumber, I mean the intersection between the edge and the face, to round it
   off?
3. To what extent do you now use guided circular saws in new installations?
4. You mentioned that there was a shortage of raw material, but, if I have
   understood correctly, you are exporting chips and even logs across the
   Pacific, to Japan. Will the shortage allow continuation of this or not?

MASON (H.C. (Carl), President and Technical Director, H.C. Mason Associates, Portland, Oregon. USA): As to whether sanding will replace the conventional surfacing, I surmise that eventually this will happen. The main reason it is not already happening is that sanding equipment cannot operate at the production rate necessary in the U.S. to maintain our low unit production costs. Also, sawing has to be extremely accurate to allow sanding.

Normally we remove about 0.020 to 0.025 inches of wood with a sander. We can remove more, but that is the standard for relatively high production. With a planer, obviously, we can take up to 0.25 inches and, while many of the mills in the U.S. today are quite accurate, well within 0.030 or 0.040 inches machine accuracy, this accuracy is still the exception and, therefore, sanding is not generally possible as yet.

In answer to your second question regarding rounding off of corners, eased edging of dimension lumber is a necessity brought about by the manual construction of homes from lumber primarily. As pre-fabrication of housing consumes more lumber in the U.S., then these requirements to avoid splinters in carpenters' hands will be reduced.

Regarding use of guided circular saws, this has become almost fundamental to the manufacturers of softwood lumber at high production rates and with reduced kerfs and relatively high accuracy. A great deal of research work has been done all over the world in this area, which has been quite well adopted in the U.S. People such as Portland Iron Works, Schurman, who are reasonably known here, I think, now Mainland, and Cancar. Most of the people in the business of manufacturing circle saw machine tools today in the U.S. use the guided circle saw with the loose collar. This means the saw is not attached to the arbor, it is left to float and it is held in the plane of sawing out at the gullet line, out at the tooth line. So, in order to reduce circle saw kerfs down to 1/8 inch or less, about 0.125 inches (3.18 mm), this has been a necessary trend, and it works well. Most of the circle saws used in the U.S. today are carbide tipped and there are some problems with that. The carbide-tipped tooth obviously lasts much longer than a filed one. To get these very narrow kerfs, we usually run about a 16 or 18 gauge saw, which is about 0.080 to 0.085 inches (2.03 to 2.18 mm). This does not have much stability and on a circle saw that averages from 20 to 32 inches (508 to 812.8 mm) in diameter, in order to maintain it in the saw cut, the guide out at the gullet line, or right out at the sawing edge is a very necessary part of the system.

Regarding exports, the U.S., as Mr. Wallace said, is very much a net importer of wood. However, the West coast forest, Douglas fir, Hemlock forest of the U.S. and Western Canada is a very remarkable one. It is an old forest, some of it still very live and vital, some of it fairly decayed, and it serves as the source of a great deal of clear vertical-grain millwork stock for most of the world. Materials leaving the West coast region are distributed throughout the U.S. and Canada, but a great quantity, not a majority, though, of this material is shipped overseas, to the United Kingdom and elsewhere in Europe, to South Africa, Australia and Japan.

The building materials that are imported are framing lumber, primarily from Canada, and some exotic woods, lauan and other tropical hardwoods from the Philippines, South America and Africa. So we are distributing out of our surplusses of high-grade material and importing lower grades of material for our housing.

WALLACE: Further to what has been said about Japan, one of the strongest influences on the western U.S. and Canada in the coming years is going to be the appetite of Japan for wood products, for both fibre and structural wood. In addition to the structural wood the country will inevitably need as soon as its housing industry recovers, Japan is in great need of fibre, and that fibre will have to be provided by the western part of North America and also by Australia and New Zealand. Another very strong influence, the impact of which is not yet being fully felt, is that cutting rights in Indonesia and Malaysia are now being made dependent on the establishment of manufacturing plants located in those countries rather than allowing the continued export of raw logs. All these will be influences on the changing marketing pattern of wood from North America.

ABEELS, P., Université de Louvain, Belgium: In view of the high level of mechanisation in harvesting evident from your presentation, I would like to ask Mr. Mason or Mr. Wallace what the level is of the value attributed to standing wood prior to harvesting.

MASON: The values of standing wood in the U.S. are increasing very rapidly. The primary cause is that about half of all the wood building products sold in the U.S. come from forests owned by the Federal government in the Northwest of the U.S. (I am currently writing a book on this, by the way, so you hit a nerve!) These resources are sold at public auction and they are sold to the highest bidder. Also, they are sold in an area where most of the mill operators do not own timber and the mill operators are a very substantial group, with hundreds of millions of dollars invested in plants and facilities. The price that is paid for timber, therefore, reflects the highest value, generally, that the end-product will bring on world markets and, accordingly, it keeps rising. This could price some of our building products out of the market. We have seen timber values double, in some cases more than double, just in the last year. The reason is that supply from the West Coast forests had always been far greater than anyone needed and supply was elastic. This is changed, and now supply is considerably reduced. And it is becoming an inelastic supply with a a rather elastic demand.

Because wood is a commodity, and the establishment of a price on only as little as 5 or 10% of the total volume can control the price in a commodity market, certainly the 50% sold by the Federal government at fixed prices establishes a very real floor for prices of building products throughout the U.S. and for some other products on the world market as well. Therefore, we do not know where the price of timber is going to stop. Those mills who own their own forests are enjoying very substantial increases in their net worth, all brought about by this floor or platform that is established by the timber sales policies of the U.S. Federal government in the West Coast. We have Douglas fir sales currently going on in the West Coast for 450 per thousand, that is board feet tree on the stump measured to Scribner Decimal C Scale. I am not equipped to convert that quickly to meters, but that is very expensive (1000 board feet is approximately equal to 2.36 m^3 - Editor). The same tree a year ago did well to sell at over $ 100, maybe $ 120.

ABEELS: What consequences do you foresee for sawmills in the progressive exploitation of the old growth and the increasing use of timber from replanted forests,
which usually results in log dimensions becoming smaller than those you showed
us in the presentation?

MASON: The forests that you saw, and most of the logs that you saw on the slides,
were virgin forests throughout the Northwest of the U.S. It is interesting that
we have many mills working on second growth timber both in the Northwest seen
in the slides and also across the south of the U.S. and the east of Canada and
the north of British Columbia. In this last area, the average log is about 7
inches (18 cm) in diameter and compares quite favourably with those from
forests in Sweden and Finland.

So, as the timber size is reduced (and the Federal government is, in fact,
placing some restrictions on the cutting of the old growth) it is really the
dead and dying, or the over-mature material, that is being marketed. Therefore,
the bulk of the saw timber reaching the U.S. mills today is relatively small
material, generally under 15 inches (38 cm). This means that you cannot get
very many 2 x 12's or wide widths out of this size of log. Consequently, the
bulk of the lumber produced is the 2 x 4 (which in net size, dry surfaced,
measures 1 1/2 x 3 1/2 inches). This has acted as a stimulant to the development of fabrication in the U.S. construction system. Instead of a variety of
wide dimensions, most structural members in a house constructed today are 2 x
4's made up into assemblies, either glue-laminated, or fabricated into trusses,
to serve the same purpose. And this is all as a result of a lack of the wide
widths available from larger trees.

ABEELS: What are the attitudes of North American sawmill proprietors regarding
the question of protection of forests and the prohibition, or reduction, of
clear felling?

MASON: If I understand, the question regards reduction in cutting. We are seeing a very severe reduction in the cutting in the U.S., primarily as it effects
public land. This is brought about by environmentalist pressure, by politics
in general. Those lands belong to the people, they are administered by the
government, and the government in the last few years has said this land is not
necessarily here to grow trees.

In industry, we are trying to work with the government on what we would
call a "best use" policy, where some lands, obviously, are not better suited to
growing forest. The very high alpine forests are very low in production, forests
that are productive we are trying to keep in production. However, there is no
question that even the productive forests will be reduced substantially from
the productive resource.

Another very serious problem is that most of the re-growth or the plantation or the reafforestation that comes about in the U.S. on Federal land occurs
primarily through happenstance. The small cut out areas are left to seed, re-
seed, and re-germinate themselves. Well, some do, some do not. It is not a
very politically expedient thing to vote money in the legislatures for projects
that will not come about for sixty years. So consequently, we are seeing a
substantial reduction in forests on the public land.

Now, on the private land, it is quite the contrary. Costs for owning and
operating forest land in the U.S. are probably close to those here. So
privately owned forest land in the U.S. is being replanted very carefully, very
intently, being managed quite well in most cases and is returning substantially
higher yields per acre than the original forest that was removed. The end
result of all this will be that, for the next 25 years, as nearly everyone in
the U.S. industry and the Forest Service knows, we will have to depend on the
private forests of the South for a great deal of the resource to support our
wood building systems.

SERRY: Can you give an explanation to account for the enormous contrast between
what we have just seen of the North American sawmill industry and the typical
middle-european industry? Could it be due to the local building system or to do
with the dimension standardization system?

MASON: Assuming the contrast Mr Serry sees is in size and capacity of mills, I
can give some answers. For one thing, the timber supply is denser and the
yield per acre is much higher, at least as far as the virgin stands are concerned.
The road systems we have are good. We are used to going from 50 to 100 miles
(80 to 160 km) from a mill for raw material, or further. With good economy
within this radius, the economy of building a mill develops around a certain
sized facility.

The sawmills with the lowest unit production cost in the U.S. produce
about 85 million board feet of lumber per year (200,600 m^3). This is related to
the American production necessary to support a given capital investment. A mill
of that type today in the U.S. would cost about $ 10 million and to build one
half that size could cost about $ 8 million, and to build one a quarter of that
size would cost about $ 6 million. This is one factor.

The other factors are the two distinct methods of marketing from the two
regions. Here, you have a more customized method of marketing, as I understand
the European system. The North American system of marketing is completely that
of a commodity: it is sold and marketed in the same way as grains and other com-
modities, sold to the highest bidder in large quantities, in large volumes, and
it has lent itself to the development of an industry that is a large industry.
And you will find that the major producers in the U.S. are not just a mill, they
are a production system involving actual house construction, finance, certainly
warehousing, transportation, all the way back to the forest. People like
Georgia Pacific, Louisiana Pacific, who are now the largest in lumber production,
Weyerhaeuser, most of the large paper companies who are in the lumber business,
all are reasonably integrated right out to the market place. Therefore, we are
dealing with a very standard commodity, 2 x 4, 2 x 6, 2 x 8, 2 x 10, and 2 x
12 - from 8 to 24 feet long. Boards, 1-inch board of 4, 6, 8, 10 and 12, 8 to
24 feet long. And about four or five grades, and that is it.

7

USA methods for production of sawngoods for construction: are they suitable for Europe?

By H. C. Mason, President of H.C. Mason & Associates, Inc., Gladstone, Oregon.

I have been asked to discuss with you today the question, "USA methods for production of sawn goods for construction: Are they suitable for Europe?" As to providing you with a definite answer to this, I am not certain it is possible. I am certain, however, that wood manufacturing techniques throughout the world will become more uniform, or standardized, because of the workings of market demand and the economics of wood cost and conversion.

I will concentrate on explaining the methods we use and the economic reasons behind them. Specifically, I would like to direct my remarks to two definitions of high output: high output in terms of wood recovery, and high output in terms of volumetric production.

The "American techniques" of lumber manufacturing have been developed historically around the philosophy of high volumetric production and low unit cost. In years past, resource was abundant and elastic. Demand was high and elastic. Labour was increasingly more expensive, and manufacturing in itself constituted one-half or more of total product cost, the balance being borne by the cost of the raw material.

Today, these conditions have changed in US and Canada to where raw material constitutes a very high percentage of total product cost, at times up to 80%. In contrast, labour constitutes a diminishing percentage of total product cost. Still other costs, such as capital, taxes, insurance, continue to increases but at a slower pace than raw material. In addition, the once vast virgin forests of the United States are being depleted. Now only sound timber management policies, on both government owned and private forest lands, are able to keep pace with the rising demands for wood products, pulp and paper.

Therefore, in light of the above, with raw material becoming very valuable and constituting an increasing percentage of product cost, the unit cost of production, while important, does not command the total attention of the American sawmill operator that it once did. In place of unit cost, the objective of increased recovery of wood products now commands a more important position in respect to design and management of the modern American sawmill. To gain a perspective on recovery, it might be helpful at this point to define in production terms the minimum level at which a US sawmill can be considered a major lumber producer. To be a major producer, a sawmill must produce at an annual rate of 60 million board feet lumber or more per year. This equals 142,000 m^3.

There are very few economically viable lumber manufacturing units producing dimension or framing lumber from softwood logs in the US below the level of 30 million board feet per year, or 71,000 m^3 per year. What we in the United States consider a large production unit, of which there are many, is a sawmill with 100 million board feet lumber production or more per year, or 250,000 m^3. But large is not the final solution. The highest relative rate of efficiency is generally found in those mills producing between 65 to 85 million board feet, or 150 to 200,000 m^3 of softwood framing lumber per year. This is because this level of operation is most often accomplished in a single facility. And single facilities have a minimum of overhead and the lowest reported unit costs. Below 65 million board feet (150,000 m^3), unit costs tend to rise -- above 85 million (200,000 m^3) unit costs tend to rise. Mills within this production range are large enough to employ the highest level of technology, can return the capital investment necessary to equip them with the most modern equipment, and still show a decreasing unit cost. Below 30 million board feet, or 71,000 m^3, it is extremely difficult to support the capital investment necessary to accomplish a high level of sophistication in milling.

Mills operating on average virgin forests in the US and Canada use logs having a minimum diameter of 12 cm and a maximum diameter of 200 cm. The average diameter falls between 40 and 100 cm. Yet, as North America's virgin forests have been harvested, the industry has turned to second growth forests for its logs. With second growth, or plantation-type logs, we are faced with milling logs with smaller diameters. It is common in the US and Canada today to operate totally with logs having small-end diameters ranging from 10 centimeters to 40 centimeters, with an average diameter at 25 cm.

In sum, US logs, other than the virgin fir, hemlock, redwood, cedar and pine of the Northwest, compare closely in size with logs milled in Germany and the Scandinavian countries. Therefore, we can conclude that European and US mills are operating on similar size resource.

The other principal factor affecting lumber production in America, and in the world, is the decreasing availability of resource and its consequent increasing value. Throughout North America, this has resulted in a great deal of attention being paid to wood recovery. With resource constituting the greatest percentage of product cost, it is extremely important that all lumber manufacturers realize the highest possible net recovery. Sawmills throughout North America experience yields from 5 1/2 board feet per cubic foot to 9 board feet per cubic foot. Our most modern and efficient sawmill facilities generally recover from 7 to 8 board feet or more per cubic foot, depending on log size.

I should explain that the nominal method of recovery, as expressed throughout North America, is now referred to as the Lumber Recovery Factor, or LRF. This factor is expressed as the amount of board feet of lumber production available from a cubic foot of log input. Also, I should state that while a substantial part of the industry still operates on log scales such as Scribner, Decimal C, Doyle, International Log Rule, and others, there is a concerted effort by our industry to adopt the cubic foot measure as a standard. Under a cubic system, the common unit of log measurement would be the cunit, or 100 cubic feet. Obviously, this is a much easier method of measurement and more accurate. In addition, it is much easier to convert it to the metric system than are any of the prevailing log scaling systems. In this respect, if I may say so parenthetically, we in the United States' wood industry are trailing our European counterparts.

I turn now to a closer look at yield, or recovery. Recovery is the product of four factors: (1) product line or the types of products that are produced, full sawn, scant, etc., (2) oversawing to accomodate variations in sawing, (3) kerf, planing, and shrinkage allowances, and (4) method of breakdown or technique employed in sawing the log into lumber products.

1. Product line is the best mix of products for a given grade size and species of log to produce the maximum return, framing versus boards versus timber versus veneer, etc.

2. Sawing variation is the permitted variation from a standard which is established as minimum to produce products acceptable in the market-place. A good example is dimension or framing lumber: In many places in the US, dimension lumber is cut to a thickness of 1 5/8 inches or 41.275 mm. Obviously, the size that can be cut depends entirely on the accuracy of the machine tools in the mill. In the final analysis, the accuracy of a mill is determined by the mill's most inaccurate major piece of machinery.

3. Machine allowance, the allowance calculated for shrinkage, planing and saw kerf, is the other mechanical variable which accounts for the balance of the loss in producing lumber to dry finish sizes. As such, it plays a proportionate part in the determination of minimum planing allowance and saw kerf are extremely important to obtaining high rates of recovery.

4. Method of sawing, which is the final point affecting wood recovery, simply involves understanding of how a log should be sawn. There is substantial difference in yield based on a log that is live sawn, half taper sawn, full taper sawn, symmetrically sawn, quarter sawn, or based on Best Opening Face techniques. The consideration of the method used to convert logs to lumber products enjoys a high priority in the US. Undoubtedly, most managers in the US industry now comprehend the advantages of cutting lumber as accurately, as simply, as scantily as possible, and are attempting to do so. This philosophy is also being applied to cutting uppers, clears, exports, and timbers. And it is these high-grade products,

which command very high prices in the marketplace and are
diminishing in supply, that the industry is now working hard
to conserve and extend by use of more efficient methods.

I move now to the subjects of recovery and grade. Recovery and grade of
lumber products is the result of the quality of the log and the type of machinery used to break the log down into lumber. The difference in break-down techniques for similar end products is profound.

Generally, live sawing, in the case of a relatively defect-free log, with
recovery the only consideration, will show the highest recovery when the log is
to be cut into dimension products. Realistically, however, we all know that knot
patterns in the center of a log, producing spike knots and cup in drying, dic-
tate cutting patterns that avoid live sawing through the heart. The point I wish
to make is that there is a substantial difference in recovery based on how you
cut.

How you cut is determined by the species and type and size of log, together with the end-product desired - all of these factors having profound
effect on recovery.

The management philosophy required to transform this type of data into
profits is the acknowledgement that every ounce of fibre saved, by virtue of
reduced kerf, reduced shrinkage, less planing allowance, less sawing variation,
or better geometry in cutting patterns, results in additional end product. In
America, as obvious as it may seem, this kind of thinking was not well understood a few years ago. But I believe it is much better understood today, and
the following example shows a typical analysis of the relationship between these factors and why this relationship has become so important.

Certainly a log 4 inches or 10 cm in diameter which has the potential to
produce only one US "2 x 4" is not affected by any of these factors. Any improvement in kerf or sawing patterns would not improve the quality of the end
product. At this diameter, the rate of return expected, due to fibre savings,
is zero. From logs with diameters greater than 10 centimeters, it is, however,
theoretically possible to recover additional end-product. A log of infinite
diameter would produce 100% of all that is cut. Therefore, if we prepare a curve relating potential return from fibre savings to log diameter, we will find
that potential return increases with increasing log diameter. We find a relatively high potential return for diameters in excess of 40 cm. We find a potential return well in excess of 50 % for logs even as small as 25 or 30 cm in diameter.

Obviously, this curve will be different for every mill. A curve for a mill
which has a relatively high recovery factor will be strikingly different from
the curve or the rate of return potential for a mill which has large kerf and
poor sawing practices. An important point to note is that all of these functions
can be modeled, can be simulated arithmetically and tested with the computer. We
have a variety of techniques for making these analyses and, I can assure you,
this can be done with a high level of reliability. This, again, is based on the
premise that a major portion of all calculated savings can be found in additional end product, or can be programmed into additional end products. These analyses presume the fact that any mill operating competitively today is run in an

orderly, business-like fashion and that the management decisions are made on the
basis of intelligent rationale. By determining the kerfs, sawing allowances, the
fibre loss, and making a definite determination of the amount of fibre being
used by a mill to develop a unit of salable product, we can apply any reasonable
hypothetical alternative with a very high degree of predictability. When this
type of analysis determines a significant potential increase in recovery, based
on either a change of product, a change in kerf, or a change in sawing practice,
we are able to make these changes and see the results in additional product. To
summarize, US production is increasingly tied to the economics of recovery as
determined by careful analysis. We have been doing this in our business now for
some eight years, applying these techniques to a wide range of mills throughout the United States, and we are frankly a little amazed at the reliability of
results obtained by this type of analysis.

Up to this point in my talk, we have looked at the four factors affecting
yield and grade. Now, I believe it might be helpful to discuss specifically some
of the machine tools being used to accomplish increased US recovery and production.

The North American sawmill industry typically makes use of the following
basic conversion units: The conventional bandmill and carriage, both single cut
and double cut, wherein we are currently equipping the carriage with precision
ball screw setworks, numerical control drives, and computer controls. This
equipment is used for logs from 2.5 m in length and longer and from diameters of
30 cm and up, where grade recovery, timbers, clears, and other considerations to
maximize dollar recovery must be considered as well as volumetric recovery.

For logs 35 cm in diameter and smaller, there are basically three general
conversion units in use in America today. The first is the round log quad band
saw. In a quad band, logs are sawn on a half taper basis with four saw lines being placed in the log simultaneously. These are designed with a 5-foot (1.5-m)
wheel diameter bandmill arranged in a four band configuration or quad. The second machine is the four-sided chipper canter. Perhaps the best known firms producing four-sided chipper canters are Canadian Car and Foundry, whose trade name
is Chip-N-Saw, and Stetson Ross Beaver. The Chip-N-Saw, Beaver and Soderhamn-
Adco West and similar machines consist of a square or a profile canter made up of
four chipping heads, coupled with either a circle saw or a band saw sawing section, so that the log is chipped and sawn simultaneously in one action. A third
breakdown machine is the two-sided chipper canter. This machine is used primarily in eastern Canada and, while a high production machine capable of infeed
speeds of over 80 m/min it does not have the recovery potential of either the
quad band or the four-sided canter saw type. The two-sided chipper canter should
by very familiar to Europeans inasmuch as it was primarily developed in Germany
and Sweden, and imported to North America, while the quad band and the four-sided
chipper canter are basically North American in origin and development.

The four-sided canter with direct coupled saws, circle or band, provides
a high level of control in pre-established sawing geometry, which adapts well to
the use of the scanner and computer for control.

Secondary mill equipment - or downstream equipment - typically consists of
shift saw edgers, circle gang edgers, and linebar resaws. The shift saw edgers,
such as those manufactured by Schurman or Portland Iron Works, have three to

five shift saws which are positioned pneumatically or hydraulically by the operator, for remote operation and have a relatively high speed production capability. A second machine found downstream is the battery, or circle saw gang, edger which may be single arbor or double arbor type, utilizing both the strobe saw and the guided saw. Depth of cut is 10 to 15 cm for the single arbor and 20 and 30 cm for the double arbor edger. In North American mills, these are replacing the framesaw so typical in European mills and still common in the US and Canada.

Perhaps the most exciting equipment downstream is the small band resaw such as the 5-foot or 1.5 meter wheel type. This piece of equipment is built in single or twin configurations. Most often a twin is used equipped with a stationary or solid line bar and two movable band saws and operated by precision ball screws, digital setworks and computerized control. As such it is able to provide most of the general functions of the band mill and carriage with a very high level of accuracy and flexibility. This piece of equipment requires the input piece to have two sides, 90 degrees apart, which have been faced by some previous method, either by a canter or by a headrig and carriage, to provide a log with a base and a straight edge so it can be milled by the twin units. Provided with a flat for conveying and one straight edge for the linebar which gives the operator a stable platform, the system can give him most of the options available to a sawyer with a bandmill and carriage.

Another major development is the computer-controlled precision ball screw carriage, which has a high level of repeatable accuracy. This piece of equipment reduces to a minimum the workload imposed on the sawyer. This is done by removing the carriage dial and substituting a digital read out. The advantage of this system is that it requires the sawyer to think only in nominals of 1-inch, 2-inch - US dimension lumber of 4, 6, 8, 10, 12 inches, etc.

Only the computer knows exactly the set decimal size and this to three decimal points. Accumulation of multiple lines, backstands with a width size cant plus any number of nominal lines are all calculated by the computer, releasing the sawyer to concentrate on log handling and the decision as to products to be cut. These carriages cost in excess of US $ 250,000, but provide sawing accuracies of a few thousands of an inch variation.

Other equipment of note currently going into mills in North America is the chipping slabber and the chipping profile edger. The slabber is an aid in removing large unmerchantable slabs and protrusions. The chipper edger accomplished the same thing that the saw edger does, except that the waste is removed by a chipper head. This allows a milling operation without the problems generally associated with edgings slabs and trim in the mill.

Additional mill equipment to consider is the conventional resaw, the sawmill gang trimmer which is used in the larger mills in North America and in Europe and, finally, the automatic lumber sorting system. Generally, the trimmers and the sorting systems in mills producing the volumes mentioned are an integral unit, with thickness, length and width being determined automatically. Lumber is delivered to the infeed of the trimmer and positioned. Once positioned, the functions for trimming, sorting, stacking and even sticking for kiln drying are accomplished automatically, without the necessity of additional personnel. These very expensive systems costing in excess of US $ 1,000,000 are providing the US lumber industry with means of reducing unskilled labour requirements - but at the same time creating a capital intensive industry.

Such systems are capable of handling lumber production well in excess of one piece per second and, in some cases, as high as eighty pieces per minute throughput.

The machine tools which have been briefly described are the foundation for the technology of the modern mill.

We should note that in most of the modern mills, those constructed within the last few years, many of the set control functions are being handled by mini-computers. With the development of accurate photo-electric sensors, accurate mechanical measuring systems and other interface equipment, the computer, together with software programming developed for optimum yield, has played a very important part in providing very high levels of pre-determined recovery, while enabling relatively high volume production. These sawing programmes are generally based on maximum value return not always compatible with maximum volumetric recovery.

The scanner computer system, coupled with the numerically controlled precision ball screw, is probably one of the major changes to have occurred in North America in the last few years to facilitate the high rates of recovery production that we currently receive. Another important development has been the general use of the high strain band saw in sawing systems wherein increased strains of 2 to 2 1/2 times those previously obtained are now employed generally throughout North America on the new machine tools. Increased strains provide greater stability, greater accuracy, and higher feed rates with a reduced kerf. For the most part, sawing systems cutting softwood throughout the United States are experiencing good results with kerfs of approximately 1/8 inch (3.175 mm) and feed rates of 45 to 80 m/min.

As you know, the high strain band mill was originally a development of Europe. It was imported to North America during the late 1960's where it is being manufactured by most major mill machinery supplies and has been well accepted by our industry.

Another development, as previously described, is the whole-log chipper canter. In its most acceptable and useful form, it is built in the form of a four-sided chipping, canting, sawing unit that in effect cuts a spline, or control guide, on the bottom of a log to direct the log through the machine. It then does all of the sawing and chipping in one function, yielding lumber of relatively high accuracy and at relatively high production rates. The whole-log chipper canter is the most common method of sawing small logs, of 35 cm in diameter and smaller, in North America today. These chipping machines are possible due to the grading and marketing practice of framing lumber in the US and Canada which requires lumber to be surfaced four sides to accurate dimensions.

Let us turn now to a few observations about the European milling methods. From numerous trips, over many years, to northern Europe, I realize that most of your sawing systems make use of frame saws, and band and circle saws, also the general concept of milling employed by most of the larger mill operations here in Europe entails a high level of log sorting and grading prior to milling. This is a concept which has been studied by US operators in great detail and, for the most part, we feel it is not applicable to most situations in North America.

The reasons are as follows: Any log sorting programme adds substantial
unit costs. In most cases, this is an additional burden which must be borne by
the raw material, driving its cost up. And this cost is not always recoverable
through the milling operation. The best concept in the US has been generally to
develop and make use of sawing systems that are flexible enough to accommodate a
relatively wide range of diameters and lengths. As an example, logs from 10 cm
to 40 cm will be accommodated in mixed order by one type of machine system which
has the ability to develop optimum rates of recovery at throughputs of 35 to
50 m/min. Moreover, such systems allow for complete flexibility in sawing pattern
and cutting design - from the smallest to the largest log - without sacrificing
any substantial loss due to the changes in the settings of the machine tool. As
an example, 5-m long logs generally would require a maximum of 2 m of gap to accommodate the set changes of an average machine (the average set time required is
no more than 1 meter of space between logs.) Therefore, if we programme an average space of 7 m for a 5-m log at 50 m/min, we are able to accommodate a total
production of 7 logs/min of a range from 10 to 40 cm, and still provide for the
highest level of sophistication in sawing design. Clearly, this system adds nothing to unit costs inherent in a log sorting system. Thus, it has allowed us to
accomplish lower manufacturing costs together with high production rates, to say
nothing of a relative simplicity of factory design and reduced capital cost.

Therefore, sorting is generally limited to distribution by size, grade and
species between small lumber logs, larger lumber logs, plywood and chips.

It should be noted at this point that designing flexible high production
machines and materials handling systems is a necessity forced upon our industry
probably more than yours. Just looking at our wide product range gives you some
idea. The North American system provides for a wide level of flexibility going
from 1-inch board products, which are finished, surfaced net 3/4 to 2-inch dimension boards used for framing and measuring net 1 1/2-inches in thickness and sold
in widths of 4, 6, 8, 10, and 12-inch and in 2-foot multiples of lengths of 8 to
24 feet heavy dimension in 3, 4, and 6-inch thicknesses and widths from 4 to
18 inches.

A variety of industrial type products, including clears and millwork stock
of all thicknesses, widths and lengths, a vast number of grades and specifications, as well as an entirely separate specification for export items many of
which are metric sizes and which are basically high grade clears and industrial
cuttings are shipped from North America to the United Kingdom, Commonwealth
countries, Europe and Japan. Most mills in North America are equipped to accommodate this flexibility in manufacturing. The machine tools employed must have this
flexibility built in. Further to this, it should be realized that softwoods in
the US and Canada generally number some six to eight basic species ranging from
Douglas fir and hemlock to some four or five basic species of pine - all of which
may grow in a single geographic area together with cedar, larch, spruce, and a
variety of true firs which make up a real potpourri of species which comprise the
forest of a good part of this area. Only in the southeastern and northeastern
part of United States and northeastern Canada we are limited generally to three
or four softwood species which are generally milled and marketed as one species,
or two groups of species.

Obviously, the complexities of the softwood lumber industry are extreme. A
sizeable portion of the better grade and larger size classifications of this re-

source are used for veneer and plywood, as most structural building in North America is accomplished with sheathing grade plywood sold in 1/2-, 5/8-, and 3/4-inch thicknesses and 4 x 8 feet for width and length. This, together with the 8-foot long American 2 x 4-inch stud and the 1/2-inch by 4 x 8-foot gypsum sheetrock comprise the basic framing and interior wall system of American housing, and constitutes the products which are developed from most of our softwood resource. In the US today, plywood must compete with domestic and export lumber markets for the larger diameter and higher grades of logs.

We have not discussed hardwood manufacturing in the US. But its manufacturing parallels that of the softwoods. It does, however, rely substantially more on a grade-selective milling procedure, and due to lower production rates, smaller size operations have not as yet made general use of the new more costly machine tools and systems described as common for softwoods.

In conclusion, it would appear that most of the machine tools used by the American lumber manufacturers are applicable for use in Europe. I would refer especially to the small log conversion units such as the Chip-N-Saw, Beaver, the quad band, the canters, the twin resaw and other equipment of this type, inasmuch as it is accurate, makes use of thin kerfs, is generally flexible, and could provide milling systems that would be versatile and could possibly result in conversion of lumber products at a lower unit cost than some systems currently in use in northern Europe; and are flexible enough to adapt to the lower production rates and requirements of the European mill. I would encourage you to investigate these systems which, undoubtedly, will be employed in different configurations than those used in the US.

Discussion

DOUAY, J., Bois Déroulés Océan, France: Regarding the term "lumber recovery factor" that he used, does this apply to volume over-bark or under-bark; secondly, I would like to know if sawmills operating in the southeastern U.S. cutting small-dimension pines have any particular problems arising from their having been used to produce naval stores, i.e. resin.

MASON: On lumber recovery factor I assume that probably most of you know that in the American system we sell a 2 x 4 net 1 1/2-inch x 3 1/2-inch, for 2-inch x 4-inch. Therefore, there is a built-in "overrun". All lumber products are generally that way. In measurement of net wood the measurements are under-bark. Recovery is calculated as the number of board feet per cubic foot produced.

Recovery is highly dependent on diameter. In the diameter range 10 to 15 cm, the possible recovery factor would be about 6 and a log 60 cm in diameter would have a possible recovery factor of nine, ten, or even eleven board feet per cubic foot. Theoretically you could not have more than twelve board feet per cubic foot, because there are 12 board feet in a cubic foot. However, due to the "scant" size, it is theoretically possible on a log of infinite diameter to have a recovery rate of nearly 16. In the South of the U.S., for example, we enjoy a recovery rate of about 7 board feet per cubic foot in the better mills, maybe 6 1/2 for small logs.

Regarding the milling of turpentine or naval stores pine, this is being done on a large scale in northern Florida, southern Alabama and southern Georgia. The only problems are the small nails and tin that are ingrown in the logs, and this just means you change saws fairly often. But it is not a particular hazard.

MOHR, K., Karl Richtberg, Fed. Rep. Germany: Mr. Mason, you spoke about the versatility of the sawing system in the United States that makes it possible to saw any diameter of timber without it being sorted. My first question is: how are you able to sort the different species and dimensions after having sawn them, and how do you saw them in terms of quality? How can you do this sorting when you have six or seven different species found in one area? Do you sort them prior to sawing or after?

MASON: This arises especially on the west coast of California and Oregon where we have seven or eight major species, some of them growing in the forest together in groups; however, most of them are identifiable and, in many cases, they are already sorted when they come out of the woods. Otherwise, the sorting will be done on arrival at the mill.

In areas of the U.S., such as the South, where we have five species of pine, all of the pines are marketed as Southern Yellow pines and the only other sorting takes place in the grade, and that has to do primarily with the openness of grain, the number of annual growth rings per inch, together with the quality.

Regarding grading, we generally have four grades in the U.S., as far as dimension is concerned: number 1, 2, 3, 4; or construction no. 1, standard, utility, economy, and then a number of clear grades, almost a whole bookfull of them. The situation becomes very complicated as far as all of the export and what we call shopgrades that go into the making of doors and windows and sash, and this type of operation.

8

The USA's first wood fibre recovery plant: Kelbro Corporation

By James C. Wallace, Vice-President, Miller Freeman Publications, San Francisco, USA, and Publisher, WORLD WOOD and FOREST INDUSTRIES magazines.

The following paper was not scheduled in the printed programme of the Seminar.

Most programme presentations today deal with wood processing plants which are manufacturing one or more products. However, this presentation takes a detour - a report on a unique plant in the United States which, while using wood as its raw material, is converting scrap wood into usable, marketable fibre. It is, in fact, adding to the usable fibre supply at a time when companies are constantly looking for new sources of fibre.

The Kelbro Corporation plant at Sacramento, California, is the only one of its kind. The company has a patent on its process for recovering usable fibre and calls its plant a "fibre reclamation" operation, not a wood products plant.

Visitors from Europe, South America and Japan have visited this plant to observe the process. And there has been a stream of visitors from pulp and paper companies and from solid waste disposal companies in the United States and Canada. David Keller, president of the company, would, I am sure, be pleased to show you around the plant if you visit northern California.

The main product is pulpable wood fibre. This is the primary purpose of the Kelbro Corporation's operation. "Kelbro" represents the two Keller brothers, David and Frank, who are the principals in the company. All of the pulpable fibre is sold to Fibreboard Corporation for use in its pulp and paper mill at Antioch, California. About 35% of total fibre processed at Antioch is reclaimed wood fibre. Kelbro now delivers 33,000 m^3 of this fibre each month to Fibreboard's pulp and paper mill.

The Kelbro operation is basically a simple one. The company collects or re-
reives deliveries of scrap wood in Sacramento and within a radius of approximately 105 km of the city. The wood, dry and green, is put through a hog and then
across vibrating screens for sizing.

The dry wood is in the form of trim ends and scrap from box factories; trim
ends from moulding and woodworking plants; trim from sawmills; broken pallets from
the area's huge food processing industry; clean scrap from residential housing
construction projects; and old wood - perhaps 50 to 80 years old in some cases -
from demolition work in older residential areas. Some of this material has attached nails, bolts, bits of steel strapping, wire, fibre rope, bits of concrete
and gravel, and assorted other debris.

Much of this material used to be burned, but strict environmental regulations today no longer allow this method of disposal. Another disposal method,
sanitary landfill, is becoming less and less available. And so the Kelbro plant
provides one answer to some waste disposal problems.

The green wood comes from veneer mills, primarily. This may be in the form
of clippings, curled, or trim. There may also be some chunks, which are lilypads
trimmed from the ends of plywood peeler logs. The green material is fed into the
plant on a separate line from the dry in-feed.

There is a dry wood end of the plant, where dry wood is dumped on the
ground, for later processing, or dumped directly into a receiving pit.

A rubber tyred Hough bucket loader maintains a steady flow of dry wood to
the receiving pit. Green wood is stored separately. There are elevated storage
bins for dry sawdust and "fines" from the process. These are sold to the cattle
industry for animal bedding.

Kelbro operates its own fleet of 27 trucks and 94 chip van trailers and
collects some of its wood scrap. Other material is delivered to the Kelbro site
by independent haulers. Some of this is sold to Kelbro. Some of it is delivered
free, for disposal purpose.

The receiving pit for dry wood has sloping sides in steel plate and is
approximately 12 m long and 6 m deep. At the bottom, there are two drag chains.
These keep the material moving and advance it to a belt conveyor. This conveyor,
slightly less than 2 m wide, moves upwards in the plant to a flotation vat.

The entire conversion operation is controlled by one man, who works from
a centrally located control room.

In the flotation vat the wood does not get soaked, for it is in the vat
only a few moments. A shaft on which are welded sections of steel pipe rotates
in the vat. These serve as an agitator or flail. They also serve to advance the
wood scrap so that it drops out on the other side. Loose nails, scrap, gravel,
and other debris - sometimes totalling up to 7 kg - fall to the bottom and are
removed later. The water in the vat also serves to reduce the dust level in the
plant.

The Kelbro plant can run dry or green wood separately, or together. Most
runs are made of a blend of dry and green.

All the wood fibre then is delivered to the Montgomery hog 75 XL-KC Eat
Rite. There is a vertical drop of approximately 5 m during the feed into the hog.
The hog has a rotor approximately 190 cm long and a cutting circle of about 91 cm.
The power unit is a 600-hp electric motor. The hog operates on a punch-and-die
principle. A series of high and low teeth is mounted on the rotor. These operate
against the material and against, or through, anvils mounted on either side of
the horizontal housing.

The flow of hogged fibre coming out of the hog drops onto a small belt conveyor about 67 cm wide. It then is delivered to a pair of vibrating screens situated just inside the corner of the building. Before the hogged fibre drops to
the screens, it passes a magnet which removes the heavier chunks of metal. In
fact, all the metal that has not fallen to the bottom of the flotation vat passes
through the hog.

At this point, the acceptable fibre (passed by the screens) is transferred
to another small belt and this carries the fibre to the waiting vans. The dry
material - sawdust and "fines" - moves through pipe above the building to the
storage bins. The "overs," those too big for use by the pulp and paper mill, are
re-introduced to the system, being dropped onto the wide conveyor belt, just be-
neath the control room. These are run through the whole process again.

Just before the fibre is dropped into the waiting vans, another magnet removes the smaller bits of metal. A small trailer sited just beyond the fibre
conveying belt, and it receives wet sawdust and fines. The trailer will later be
towed away and the wet material spread on the Kelbro site to cure for 20 to 24
months before being sold to nursery (plant and florist) operators.

All van loading is done in tandem fashion and deliveries are in tandem
hauls. A Kelbro-designed load leveler is mounted at the loading station. As fibre piles up in the van and begins to pyramid, a series of horizontal metal bars,
spanning the van and mounted on an endless mount, scrapes the top and gradually
builds the load of chips forward in the van. When the lead van is full, it is
moved up and the second unit is loaded. Truck and trailer combinations may make
four to six trips daily to the pulp and paper plant, some 95 km from Sacramento.

Along with the intake of dry and green wood scrap, the mill holds its dry
sawdust in a 6-m^3 bin. Green sawdust is spread out to cure and dry before going
into the bin.

Another "product" produced here is manzanita (Arctostaphylos species), a
brush species that occupies much commercial forest land in the California Sierra
Nevada pine stands where it is an obstacle to commercial forest management. Kelbro harvests chunks of this tough, dense wood and cures it for some months in
the yard. Then the chunks are put through the hog. The resulting fibre is bagged
and sold to supermarkets as a barbecue fuel called "Manzaquets." It burns hot
and fast and leaves no ashes.

So the total list of products is impressive, and all from scrap material :
pulpable fibre, material for bedding animals, material for nurseries, metal to
scrap metal dealers, and a barbecue fuel to the public.

Discussion

DOUAY, J., Bois Déroulés Océan, France: What type of knife or metal parts are used in these hoggers that they are able to work on this scrap wood so full of metal, and how often do the movable parts have to be changed?

MASON: The machine is a relatively conventional hog made by the Jacksonville Blow Pipe Co, called the Montgomery. The anvils and knives are hard-surfaced and, depending on the amount of scrap metal that is being put through the machine, you have a proportionate amount of maintenance. When you do replace a knife or anvil and the cutter, and this can be fairly often, there is a considerable amount of downtime.

9

Cutting for maximum value using computer programming techniques

Part I

By Herr Ing. Hans Pliessnig, Director,
Pliessnig Company, Klagenfurt, Austria.

First of all, it would appear advisable to define the term "sawing programme". Applied to installations primarily concerned with the shaping of long roundwood for production of telegraph poles and with the production of building timber and sawntimber according to sawing schedules, it means a programme for the optimum cutting up of stems into parts, taking the prevailing market situation into consideration.

In sawmills where the main product is sawntimber of various dimensions, the term "rationalizing the sawing programme" is understood as a procedure which enables the timber, generally stored in the form of ready-cut logs, to be converted into sawntimber in such a way that it brings in the highest possible return in the prevailing market situation.

This paper will first deal with equipment for the optimum utilization of the stem by cross-cutting. This will be followed by a short review of pre-calculation and rationalization of the yield in bandmill and framesaw mills. Finally, a plan will be put forward for discussion which comprises the rationalization of stem utilisation and yield and the control of the results obtained, and which includes a self-correcting and self-improving overall system governed by a closed control electronic circuit.

The object of cross-cutting for optimum wood utilization is to cut each trunk into sections in such a way that, taking advantage of the prevailing market conditions, the largest possible quantity of those lines which command the highest prices is produced.

Since prior cross-cutting, at the felling site, for instance, restricts the number of possible ultimate cross-cutting combinations, roundwood should be delivered to the sawmill in full lengths.

Round timber is a natural product. It differs from many other raw materials in that its shape often deviates considerably from geometrical forms.

In order to be able to decide on the optimum cross-cutting of a stem one must first know its shape accurately. The traditional way of manual measurement in the forest of the length and middle diameter of a trunk may provide a good enough measure for the sale of the timber but this simple method of measurement certainly cannot be used as a sound basis for cross-cutting.

Attempts have been made to find whether it is possible to obtain at least an approximate picture of the general shape of a trunk by means of just a few measurements. It was discovered that the breast-height diameter (dbh) combined with its total length gave the best results. Recently an equation was published, based on the dbh and total length of a trunk, which permitted certain conclusions to be drawn with respect to the degree of tapering of the trunk – which differs from tree to tree – of North American conifers. This equation, which in principle could be adapted to the characteristics of trees in other regions, might be of some use in timber yards which receive their wood only from one specific growth area. But since the tapering of the trunks is much more pronounced in mountainous country than at lower altitudes, in my opinion such equations would serve little purpose in timber yards that receive part of their supplies from mountainous regions. For the optimum evaluation of incoming rounwood every stem must be dealt with individually.

The necessity of handling every single trunk individually means that even an experienced man cannot deal with more than about 3 stems per minute. This man has to take a large number of factors into consideration simultaneously: he must, for instance, correctly estimate the diameter of stems of many different lengths; have the market value of different sizes of sawn timber in mind; before the trunk is cut up, mentally work out the various possibilities of sectioning it, etc. It is clear even from this superficial analysis that rapidly performed measurements of as many diameters as possible along the trunk and their storage together with the length information can be ideally carried out by electronic devices, with their enormous operating speed and their storage and comparison facilities.

Modern electronic equipment has an "eye" in the form of a sensor which can record a hundred thousand and more cross-sections in a second, together with the corresponding length information. It also possesses a "brain" in the form of a mini-computer which is capable of performing a hundred thousand and more comparisons and decisions in a second. The following simplified diagram (Fig. 1) shows how a device for measuring diameter and length operates in conjunction with a mini-computer to provide optimum round timber utilisation.

Stems arrive via the debarking machine at the measuring conveyor F1/F2. They are cross-cut on the conveyor F2/F3. The feed rate, is determined by the mechanical speed of the cutting equipment, which consists of the cross-cut saw, 9, and the sectioning conveyor F2/F3.

Data on prices and the limits for the quantities of predetermined sizes of sawn timber are previously fed into the computer and stored. The computer receives accurate data concerning the shape of the trunk from the measuring unit. Moreover, the operator is provided with a simple pushbutton keyboard with which

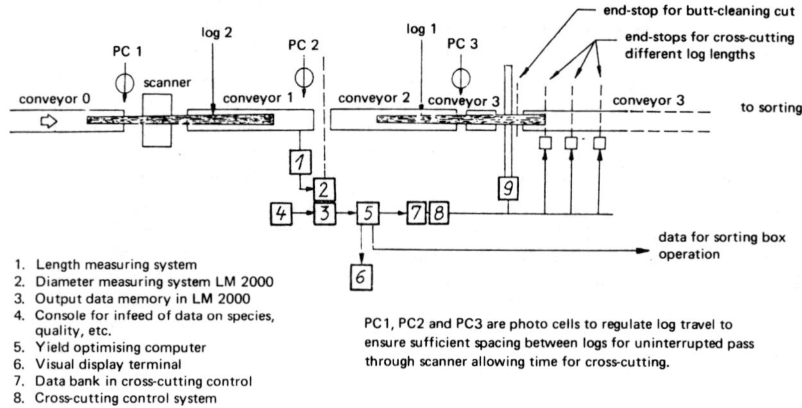

Fig. 1 - Layout and mode of operation of cross-cutting for logs,
using yield computer and electronic scanner for scaling

he can feed in additional data such as quality of the log, any rotten parts, and crookedness. On the basis of these data the computer calculates the sectioning plan which will furnish the highest yield. This cross-cutting plan can be indicated by means of a visual display.

If the operator agrees with the indicated plan, he presses a button and cross-cutting proceeds automatically. If he wants to alter the indicated plan, he can carry out cross-cutting partly or entirely by hand. The information from the computer is signalled to the individual length stops via the cutting-control system for setting the precalculated lengths of the individual sections. As an alternative to length stops, an electronic device for measuring the length can be provided, which controls the drive motor of the conveyor F3 so that the conveyor stops at the right moment for cutting off individual sections. The cross-cutting saw is likewise set in motion by information from the computer.

For inventory purposes the middle diameter and the length of the trunks can be punched on tape or written with a typewriter. The volume of each trunk can be calculated, totalled and registered by pushing a button.

Rapid and efficient computers have been on the market for years. The main progress in the processing control sector in recent years has been the development of reasonably priced mini-computers and the introduction of simplified programming methods. The pioneering work to be done in the development of new automation installations is more in the direction of data acquisition. In cross-cutting, the function of the data acquisition equipment is to register the contours of the stem accurately. It divides each trunk into sections of, normally, about 5 to 50 cm in length and then determines the smallest diameter of each of

these sections. This is important, for a decision as to optimum cutting up of
the trunk can only be taken if the actual smallest diameter of each section
is known. For optimum cutting up a measuring device is required which would
be capable of registering changes in cross-section accurately even if they occur for only a few millimetres in the length of the trunk, as is the case in
constrictions with an abrupt change of diameter, in saw-cut damage to the stem
and the like.

In order to register tapers which occupy only a short length of the trunk,
the measuring device must have a narrow field of view and be capable of carrying out a minimum number of diameter measurements per unit of time.

An example will demonstrate the difficulties that have to be overcome:
a trunk feed rate of 60 m/min, a constriction on the trunk with a length
of 4 mm is visible to its full extent in the 3 mm wide field of view of the
measuring device for only about 1 millisecond. Since it cannot be assumed that
a new measuring cycle will commence exactly at the moment when the constriction
enters the field of view of the measuring head, the measuring device must make
at least 2 complete diameter measurements in the above-mentioned thousandth of
a second.

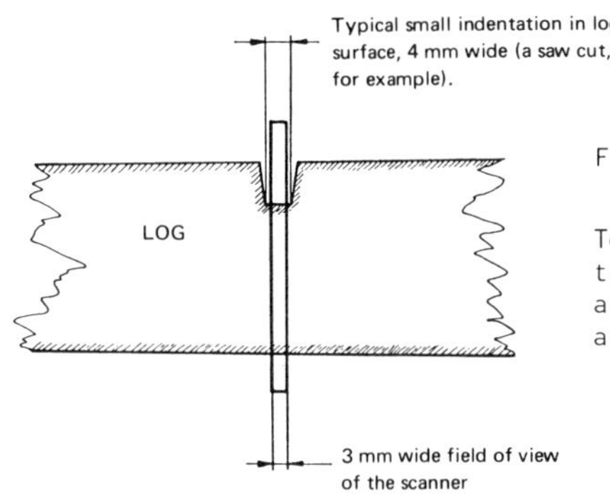

Fig. 2 -

To be able to register short-length defects
the scanner must have a narrow field of view
and be capable of carrying out scannings at
a sufficiently high rate

This short example (see Fig. 2) shows that a measuring device which "sees"
just 3 mm of a trunk moving at the rate of 60 m/min can only register a constriction of the cross-section extending 4 mm along the trunk if 2,000 or more complete diameter measurements are made per second and the smallest of these 2,000
diameters is selected as the minimum diameter. Such a high measuring sequence
can only be achieved by measuring equipment which operates without a rotating
mirror or similar devices, that is to say, completely electronically. The
required narrow field of view is also only possible with fully electronic equipment.

The more time that is required for a single diameter measurement, the
higher is the threshold for the length of cross-section deviation which it is
possible to register.

This problem can also be considered from another point of view. Since
the stem is travelling lengthwise during the measurement, the measurement is
not made exactly perpendicular to the trunk. The error arising from an inclined measuring path compared to a path at a right angle to the trunk increases,
as can be seen from fig. 3, with the duration of the measurement, i.e. with a
lower number of measurements made per second by the device employed.

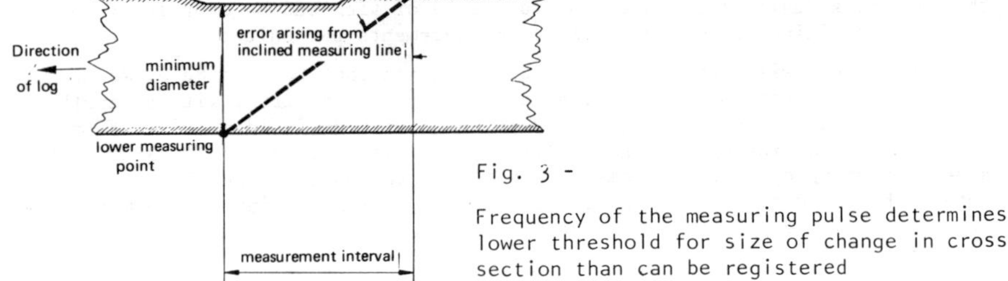

Fig. 3 -

Frequency of the measuring pulse determines
lower threshold for size of change in cross-
section than can be registered

Diameter measuring equipment which makes use of mechanical parts such as
rotating mirrors or similar devices cannot accomplish more than about 100 measurements per second. Fully electronic measuring equipment can carry out between 10 and 100,000 or more measurements per second. The sophistication and
price of such equipment increases, of course, with speed of measurement and
diminution of the field of view, since the less time and light there is available for the measurement, the more difficult it is to attain a given response
sensitivity.

To keep the measurement from being influenced by factors such as protruding
pieces of bark, knots, and the like, the measuring device must give an accuracy
for every single measurement, that can be relied on. Only then is it possible
to correctly register the smallest diameter of each section of the trunk through
automatic decisions based on suitable models. The new fully electronic measuring systems no longer need "axisparallel" light. Thus their optical system
is shielded against dirt to a large degree. This makes it possible to locate
the measuring system directly on the debarking machine. This means, among other
things, that the necessary conveyor equipment can be greatly shortened, with an
accompanying reduction in its cost. An investigation of different measuring
systems and their suitability for recording the actual smallest diameter of a
trunk will be found in "Holzrundschau" October and November 1973 and in "Holz-
Zentralblatt" of 23 November 1973 (both German - language magazines - Ed.)

Can a computer make better decisions on cross-cutting than a human operative?

In order to answer this question we must, at least briefly, examine how
an optimum sectioning plan is made. For simplicity we shall confine this examination to explaining the procedure when, for instance, a trunk 9 m long is
to be divided into two sections with one cut. Lengths 3, 4, 5 and 6 m are
permissible for the sections thus produced. In the present situation of the

market 3-m sections give a financial return of 3 units, 4-m sections 6 units,
5-m sections 8 units and 6-m sections 10 units.

In this extremely simplified example it is obvious that the price obtainable becomes more interesting in proportion to the length of the sections. One
will, therefore, try to avoid the short (3 m) sections. The optimum solution
is to divide the 9-m trunk into a 5-m section at its thick end and a 4-m section
at its thinner end.

This intentionally simple example shows that the optimum "linear" solution
(linear because the relationships can be expressed by linear equations) is mathematically a peak value proposition that is stated in the form of a function of
several variables under secondary conditions. (Anybody who is interested in
the techniques of rationalization can find an introduction to the solution of
such problems in the books "Lineare Programmierung und Erweiterungen", published
by Teubner.).

In practice, unfortunately, more than one cut is usually needed to divide
up a trunk. It must be taken into consideration that sawing off each section
influences the sectioning of the rest of the trunk. The peak value proposition
to be solved thus consists in finding, simultaneously with exhausting all the
variation possibilities for each individual section, an overall solution for
the optimum utilization of the complete trunk. Methods of solving such peak value problems are given in the book "Einführungskurs in die dynamische Programmierung", published by Springer. Extensive calculating operations are necessary
for the synthesis of this optimum utilization plan. Since in "real time" operation only a few seconds are available for performing these calculations, they
must be done at a tremendous speed, requiring use of a computer. Moreover, if
the market situation changes it must be possible to adapt the calculation programme quickly to the new situation without a complicated procedure. This requirement is also excellently fulfilled by a computer. In addition, due to its
high speed of operation and its storage facilities, the computer can take account of a very much greater number of measuring points and secondary conditions
in working out an optimum solution than could be dealt with by a human being.
This accurate adaptation to the sectioning plan for the trunk and to a given
market situation results in a much improved utilisation of the trunk.

Economic considerations for computer systems

The type and the price of a computer for use in a sectioning installation
depends on several factors. It is assumed that measuring equipment capable of
determining the smallest diameter per section, despite irregularities, is available; also, that a programme in machine language for the computer is already
available. In this case the price for a suitable optimum-caculating mini-computer, complete with input keyboard and control desk typewriter can be put at
ASch. 250,000 to 400,000 (or some US$ 11,000 to 17,000).

The price of the equipment for measuring the diameter and length of the
sections, including an interface to the computer and installation of controls
for the cutting machinery, will be, with fully electronic equipment, between
ASch. 350,000 and ASch. 800,000 (US$ 15,000 and US$ 34,600 respectively), depending on the required resolution of the measurements and the size of the smallest taper to be registered. The price for the cross-cutting equipment depends

very much on the required throughput and is not significantly higher than the
cost of manually operated sawing machinery.

The first effect of rationalization that can be expected is a distinctly
increased throughput, since manual measurements and the comparison of different
possible log lengths are eliminated. Throughput is then only limited by the efficiency of the mechanical equipment. In spite of this increased throughput
only one man is required for cross-cutting, and he has much less work to do.
In addition, a considerably higher yield can be obtained than in the case of
manual operation. Increases of yield of up to 17% have been reported.

Sawing programmes in the mill

In the production of sawntimber, too, raw material costs are much higher
than labour costs. An accurate calculation of potential yield as a basis for
increasing yield and for the establishment of a series of sawing programmes is,
therefore, equally important.

In the endeavour to obtain an optimum yield everything depends on determining the most practical way of producing given sizes and quantities of sawntimber
from the diameters, lengths, and qualities of the round timber in the log yard.
It is also important to know the ideal setting of the framesaw blades for cutting
logs classified according to smallest diameter and degree of tapering, and to
know which diameters of round timber it is best to buy. These data are assembled
on the basis of a given price per cubic metre dependent on diameter in order to
obtain the optimum yield of sawn timber to satisfy a given demand.

A rule derived from peak value calculations has been used for a long time
to determine the contours of the main product. This rule states that a rectangle
of a certain size will contain the maximum useful area when it is made into a
square. As soon as one proceeds from the smallest trunk diameter contained
within a square in order to take the secondary product into consideration, the
relationships quickly become more complicated. When one looks at a simple cross-
section it is at once apparent that a greater board area of the secondary product
can be obtained if the boards are sawn with decreasing thickness towards the
edge of the trunk, as is shown in Fig 4.

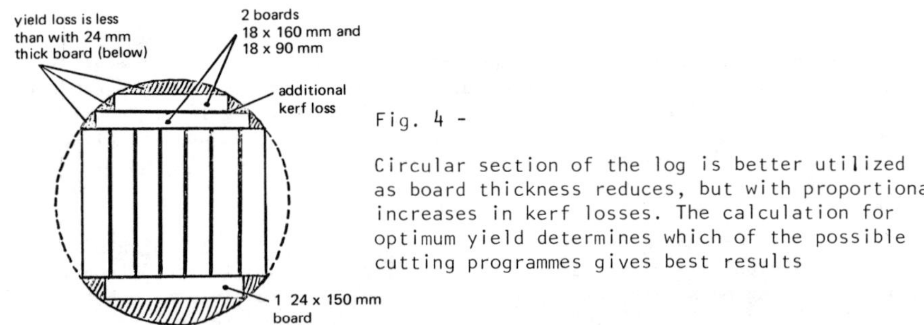

Fig. 4 -

Circular section of the log is better utilized
as board thickness reduces, but with proportional
increases in kerf losses. The calculation for
optimum yield determines which of the possible
cutting programmes gives best results

But a closer look shows that more kerfs are thereby produced and thus more
sawdust, and it is necessary to check that the extra profit obtained from the

production of two thin boards instead of one thick board is not cancelled out
by the increased number of kerfs, or even turned into a loss. It is clear,
therefore, that a peak value problem with secondary conditions is also presented
in this case. Secondary conditions are given, for instance, when boards may
not be sawn below a certain minimum specified thickness.

n The problems of optimum yield become still more complicated when the taper
of the trunk must also be taken into consideration for the purpose of calculating the optimum processing of the secondary product. It is obvious that a computer is the most suitable instrument for performing these relatively extensive
calculations. An article on the use of a computer for precalculation of the
yield can be found in "Holz-Zentralblatt" 22 February 1971. This article states
that an increased yield of an average of 3.1% in the principal product alone
can be obtained by the use of a computer.

The more factors that are included in the rationalization process, the
more satisfactory, naturally, is the result. Thus the rationalization should
be taken all the way to timber purchase with careful estimation of the dimensions, degree of tapering, of stems before felling. Quantitative yield obtainable increases with trunk diameter and decreases with increases in degree of
tapering and/or crookedness, while labour costs per cubic metre decrease with
increasing trunk diameter.

In the ideal case rationalization should then continue in the market-
oriented cross-cutting of stems. In framesaw mills the first step should be
to sort the logs according to their smallest diameter (which need by no means
be the smallest diameter of the top end if, for instance, constrictions or saw-
cuts are present), and the degree of tapering. Grading the diameter groups as
closely as possible facilitates optimum matching of the logs to the framesaw
spacing, particularly important in the case of a prism cut. It also limits the
maximum width of the side-yield.

In the newest automatic system, crooked trunks are dealt with in the following way: the operator in charge of sorting presses a "crookedness" button
which informs the automatic measuring and sorting system to down-grade the trunk
into the next lower diameter group, so that the crooked section is sawn, e.g.
in the case of prism cut, by the framesaw assigned to cutting this smaller diameter group. The sorting programme is normally stored on a plug-in circuit
card. This can be exchanged in a moment for another card with a different programme.

In automated bandsaw mills, instead of this sorting, the bandsaws are generally controlled by commands which the computer calculates on the basis of the
trunk diameters.

Checking the results

I see that what I have said up till now has met with some approval.
But what the competitive sawmill proprietor needs most of all is practical success. Therefore any rationalization of the process of cutting timber would be
incomplete without some means of checking the results obtained. In ideal cases
this check would provide information each evening regarding the yield obtained
on that same day.

A scheme for a daily check of yield, again depending on the facilities offered by a computer, is already in existence, and I would like to explain briefly how it works with the aid of the following table:

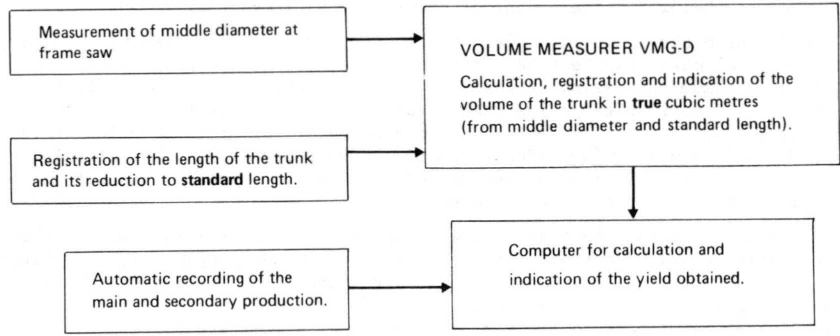

Fig. 5 - Monitoring the yield in a framesaw mill
An accurate registration of the yield provides daily information on results and enables sources of loss to be eliminated promptly

The volume measurer, which must also be used for yield monitoring, must determine the cubic content very accurately, since in difficult times a difference of a few per cent in yield can mean the difference between a profit and a loss. A precondition for accurate cubic content determination is that the measurement is made in true cubic metres (from the <u>standard</u> length of the trunk and the <u>middle</u> diameter). The yield computer calculates the cubic metre contents of the various products, adds them up, and compares the total with the consumption of round timber.

Future developments

The tendency is undoubtedly towards the replacement of decisions that are mostly made intuitively by methods that can be taught and learnt. The means to do so are already provided by the rapidly advancing technology of modern data acquisition and processing equipment. The following table (Fig 6) shows how optimum sawmill operation with a closed control circuit can be put into practice with equipment which is available today, the control circuit being responsible for optimum sectioning and yield as well as for checking the yield obtained. An important point is that, through the feedback of the results and the conclusions derived from them, a quasi-automatic correction of the whole system (especially through inclusion of these results in the data used for purchasing) can be attained.

(Note: Discussion appears on page 92 following the second half of this joint presentation).

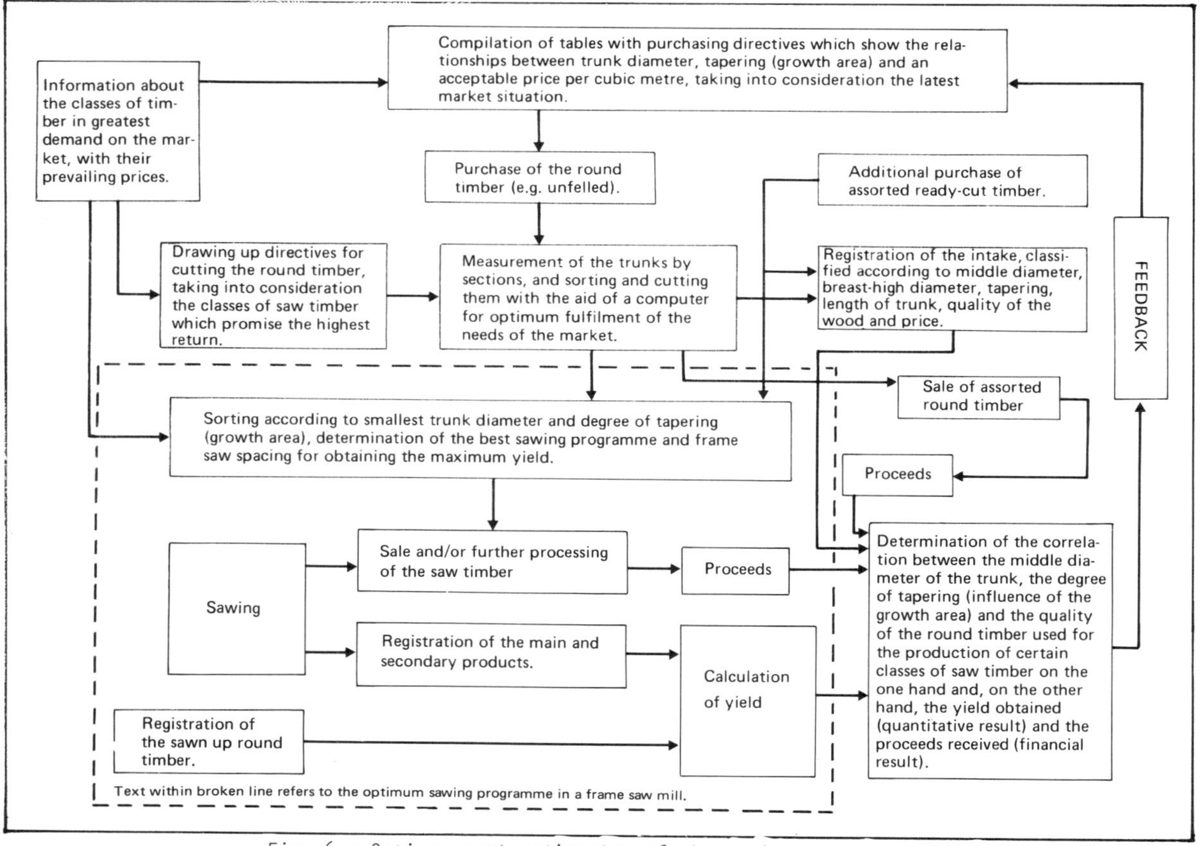

Fig. 6 - Optimum rationalisation of the sawing programme with a closed control circuit (schematic representation)

9

Cutting for maximum value using computer programming techniques

Part II

By Herr Ing. Ernst Sondermann,
Siemens AG, Erlangen, Fed. Rep. Germany.

You will all be familiar with the traditional open-air log yard where the logs are measured by hand. With the constantly increasing cost of labour and the dependence of outdoor work on the weather, a more rational procedure with improved working conditions became urgently necessary. And so, a few years ago, the mechanised log yard controlled by contacts and relays was introduced.

Recent years have witnessed enormous strides in the application of electronic systems, with their wear-proof integrated circuits. These modules and the corresponding sensors have enabled automation and data processing to be extended to a degree which was impossible with contacts and relays on account of their limited speed of operation and their relatively bulky size. The diagram in Fig 1 shows location of the sensors for electronic control in a log yard.

For the measurement of length, a vibration-proof angular stepping sensor is coupled with the conveyor belt or chain. This device functions without contacts or wearing parts. Provided that there is no slipping between the trunk and the conveyor, the number of pulses delivered by the stepping sensor per revolution determines a certain transport path.

For measuring diameters an optical-electronic light barrier with digital measurement output is used. The size of the darkened area provides a direct measurement of the diameter. The diameter of the trunk is measured as it passes the light barrier. The measurement is performed in 10 milliseconds.

The commands are given by signal transmitters and light barriers which operate without contacts and without touching the timber. When a metal object passes these devices they emit a predetermined switching signal which can be processed directly by the electronic control equipment. For example, certain functions can be triggered by parts of a machine such as the dogs on a conveyor chain as they pass by.

Processing of measurement data, control and computer units

Standard modules from the SIMATIC C3 System are employed for processing the measurement data and signals. This system operates with integrated circuits in TTL technique, such as are used in normal data processing equipment. The controls operates with synchronized pulses in order to safeguard them against internal faults. For the controls used in the wood industries the clock frequency is approximately 1 MHz. Only in this way is it possible to carry out complex calculations, such as are required for determining volume, with sufficient speed.

These plug-in standard modules with printed circuits, are known as "prints" (see view 20 on final pages of this volume). These plug-in units are accommodated in a cabinet rack (see view 21) and connected up to form functional units, the connections being made by means of the wire-wrap technique. Wiring is carried out by the SIPOMAT automatic wiring machine, resulting in a considerable saving of time in assembly and testing.

Standard functional units

Since mechanized log yards exist in various degrees of development, different standard functional units are available in the LIGNOTRON System for the wood-working industry. These units can be combined in different ways to suit individual requirements. These functional units serve the following purposes: measurement of length, measurement of volume, and positioning.

Measurement of length

From the experience gained with existing and tested installations it has been found that the best lay-out of measuring equipment for sawmills is as follows: a light barrier is located at the point where the round timber is put on the conveyor which takes the trunks to the saw. In addition, an angular stepping sensor is coupled to the conveyor chain in such a way that it delivers a pulse for every centimetre of the transportation path.

As soon as a trunk enters the light barrier, its distance from the cross-cut saw is registered (in cm) by a counter. When, on further transport of the trunk to the cross-cut saw or positioning, its end clears the light barrier, the distance registered by the counter from the light barrier to the cutting-off point is diminished through the pulses of the angular stepping sensor (1 pulse = 1 cm) by the distance which the trunk still has to travel to the cross-cut saw. Now the counter has registered and displays the length of the trunk on the control console.

Only one light barrier is needed if the distance between the point where the trunk is placed on the conveyor and the cutting-off saw is longer than or the same as the maximum length of trunk which is handled. But the shorter the distance between the point where the trunk is placed on the conveyor and the cross-cut saw, and the longer the maximum trunk length, the more light barriers will be required.

Measurement of volume

The measurement and calculation of volume begins as soon as the trunk enters the light barrier. The length of the trunk is measured by the angular stepping sensor which is driven by the conveyor. The diameter of the trunk is measured every 10 cm as it passes through the digital light barrier. Experience has shown that a very accurate volume measurement is obtained if five of these successive diameter measurements made at intervals of 10 cm are added together and the average diameter is then taken for calculating the volume of that section of the trunk. This eliminates such irregularities as knots, and minor taper, to a great extent. The total volume is then determined by the addition of these 50 cm sections, plus the remaining end section.

Positioning the log

For accurate positioning the cross-cut saw feed conveyor must be switched from normal to slow speed so the log arrives at the correct point for cross-cutting. These individual points can be adjusted by dials on the console. In this way accurate positioning is attained, with allowance for different co-efficients of friction and saw-blade widths from one installation to another.

Sorting chain loading

This unit ensures that the trunk is placed on the conveyor at the right moment, so that it is always centrally positioned, and is discharged correctly into the sorting box selected by the operator (all boxes are at a uniform distance). It is also possible to use a belt conveyor with an accurate discharge into boxes at differing distances.

Controls and displays

Operation of a mechanized and electronically automated log yard is carried out from a central console. It is provided with certain display panels in combination with the sorting system. Displays are for:

1. Trunk measurements. Indication of the different diameters of the trunk at intervals of 2 m from the stump. There is also a luminous indication of the total length of the uncut trunk at intervals of 50 cm. This is of great assistance to the sorter in dividing up the logs. Sorting on the basis of definite measurements is also more accurate than the former method of estimation by eye

2. Subdivision of the trunk. This panel is provided with a multiple subdividing field which enables the trunk to be cut into a maximum of 6 sections, with the appropriate box numbers. This subdivision of the trunk is made by selecting the corresponding section (e.g. button A) and pressing a pushbutton on the ten button keyboard to select the length of the section in cm and the corresponding box number. On pressing another button, the trunk is cut into the preselected sections. This panel also has a 4-digit numerical indicator of "length before the saw" and "length after the saw". This indication also functions if the installation is switched to manual operation.

3. For recording the production data: number of logs; length; volume, mechanical counters are used, which retain their read-out even in the event of power failure. These counters are also provided with outlets for printers or data processing equipment.

The use of these electronic controls in mechanized log yards enables a
considerable increase of through-put to be achieved, together with a saving of
labour. The non-wearing components, needing practically no maintenance, are
therefore very suitable for increasing production in log yards.

A large number of installations embodying our SIMATIC C3 modules described
above have been supplied and are in successful operation. A complete installation for a log yard which comprises all the functional units which we have
described is shown in fig.1. Nevertheless, the equipment is very compact.
The use of contacts and relays for the same purpose would have meant an unac-
ceptably high outlay.

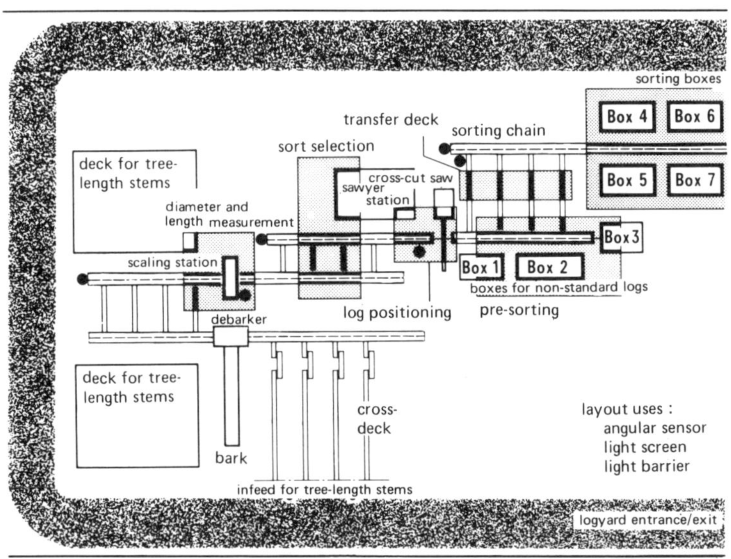

Fig. 1 - The "electronic" log yard

Discussion

FRONIUS, Programme Director: The problem of optimization is one we consider very
important and we have even tried to take it beyond the initial stage with the
whole stand, that is, to the actual filling of orders, which should also be
programmed. This would allow logs to be selected in view of how they could best
be cut and what type of product they would best make. The problem is that not
only are length, diameter, crookedness, and other properties, very important,
but also the question of branches and whether these are dead branches or not.
Perhaps Herr Pliessnig can tell us something about this.

The main problem here is programming. The initial step has been resolved by
Dr. Glück in Vienna. We here in Munich in Professor Löffler's Institute are
working primarily on the quantifying of the quality of timber, which has not so
far been taken into account. Until now, sorting has been by dimensions and only
rough estimates of quality have been made. We are endeavouring to establish a
list of faults and defects and, by different reduction factors, to arrive at a
system that would allow quality to be taken into account in the programme.

QUESTION: Is the type of equipment referred to by Herr Pliessnig and Herr
Sondermann the same or not?

PLIESSNIG: There is a considerable difference between the two types of equipment.
The Siemens installations use rotating mirrors and axis-parallel light. Our
machines do not require axis-parallel light and we have a fully electronic
system unaffected by stray particles such as saw-dust. The system, for which
patents have already been filed, works with an optical system that refracts the
image of the stem with a lense system.

SONDERMANN: I wanted to come back to the point Herr Pliessnig has just made.
In the Siemens system we use a rotating mirror. It makes one hundred measurements per second, as Herr Pliessnig has said, but we immediately enter what, in
my opinion, is a philosophical question: if the value is going to have to be
calculated afterwards, what is the purpose of doing a pre-calculation during
the actual measuring, since this operation also adds to the cost? This is why
we will continue to use the rotating mirror concept.

10

Two ways to increase product value: stress-grading and finger jointing

By Victor Serry, F.I.W.Sc., Managing Director,
Measuring and Process Control Ltd., Rainham, Essex, United Kingdom

The recent sharp increase in timber prices and the stress thrown onto the European economy by the oil crisis have focussed attention on two ways of avoiding waste in the use of wood and of improving its value as a marketable structural material.

Finger-jointing

The technique of finger-jointing arose during World War II in Germany but has been remarkably slow in winning widespread acceptance, considering that it overcomes one of the most important imperfections of timber as a raw material, the biological length limit. Finger-jointing not only reduces length waste (one of the principal causes of timber waste) but also eliminates ugly or weakening defects.

Finger-jointing has reached an important stage in its development; it is established in the technical sense but awaits wide-scale commercial exploitation. This will come about only when such processes have been shown to be economic for the individual timber processor.

In September 1971, the then Forest Products Research Laboratory, Princes Risborough, published an economic evaluation of finger-jointing processes operating in three ways:

1. To produce timber of any desired length from the available stock lengths.

2. To minimise waste by jointing short lengths and offcuts to produce usable lengths.

3. To up-grade poor quality material by removing defects and end-jointing the

remaining up-graded timber.

Plant costs were sub-divided into (a) fixed costs independent of the level of output (depreciation, interest charges and other costs such as factory services allocated usually as a result of management decisions); and (b) costs varying in proportion to the output level (direct labour cost, power consumed, machine maintenance, cutter maintenance, cost of adhesive, and cost of wood lost per joint).

The second aspect of the evaluation was to estimate the plant performance in relation to the continuous output capability of the plant and the level of plant utilisation to be expected.

The trend in finger-jointing plant development is for an increasing proportion of the operating cycle to use sequence or automatic control, and this makes it easier to estimate the continuous output capability. The level of plant utilisation took into account a number of factors including allowances for machines breakdown, routine maintenance and cleaning down, and setting up of new cutters. The following table shows a sample of results for regrading and reclamation operations as well as the production of particular components.

Operation type	Input length mm	Output length mm	Through-put rate	No. of operators	Average batch size	Machining assembly & adhesive cost/joint
Regrade *	2,750x150x50	2,100x50 door jambs	0.5 m/s 100 ft/min	12	550	16p 0.95DMk.
Reclamation	900x150x50 offcuts	1,800x150x50 studs	0.85 m/s 17 ft/min	3	500	9p 0.53DMk.
Reclamation	900x100x50 offcuts	2,400x100x50 posts	0.055 m/s 11 ft/min	3	1,000	8p 0.48DMk.
Component	random lengths 150x50	1,800x150x50 studs	0.075 m/s 15 ft/min	4	500	11p 0.65DMk.
Component	5,100x100x38	7,745x100x38 beams	0.13 m/s 26 ft/min	3	122	15p 0.89DMk.
Component	random lengths 125x50	2,430x125x50 beams	0.035 m/s 7 ft/min	4	40	25p 1.5DMk.
	random lengths 150x38	9,600x150x38	0.07 m/s 14 ft/min	4	250	27p 1.6DMk.

* 5machinists & 7 labourers & graders continuous operation

Given such figures it is relatively easy to estimate the overall economics of any particular operation by comparing the costs of finger-jointing with either the increased revenue or reduced costs whichever is appropriate to the operator concerned.

The reason for the slow adoption of finger-jointing must be sought in the
low and stable price of timber relative to its utility, over many years. Now
that timber has ceased to be under valued, the advantages of finger-jointing in
terms of cost saving have doubled or trebled, while the costs of the process itself have risen only slightly, to between 10 and 20p (0.6 - 1.2DMk.) per joint
for most applications, or about 32p (1.9DMk.) per joint for the kind of application shown by the last two items in the table having costs of 25p (1.5DMk.) and
27p (1.6DMk.) in 1971. This is in fact far less an increase proportionately
than there has been during that time in, for example, labour costs. The biggest
change is that throughput has increased to about 7 m/min - 2 to 3 times faster.
For these reasons many companies are installing plant now, particularly in association with the manufacture of glued laminated timber structures.

On the technical aspects, it is interesting to note traditional British
practice favours a joint-length of 50 mm as against the much shorter length of
20 mm generally preferred on the continent of Europe. The latest evidence seems
to show that there is no particular advantage in joints more than 20 mm long and
it seems likely that in new installations Britain will fall into line with Europe.

Stress grading

The other powerful method we have of increasing the value of timber is by
reducing the very great uncertainty in its strength. Timber varies greatly in
strength from piece to piece and it has become an accepted part of building practice in all countries to make a large allowance for this uncertainty; this is
expensive because much more material has to be used to reduce the uncertainty
level.

Traditional methods of grading structural timber have not prevented the
wasteful use of wood because they are based on appearance, not strength, and this
in turn has led wood research institutes in consumer countries to create many individual systems of strength selection for constructional wood; this is unfortu-
nate as the real need is for a unified system which will make the marketing and
use of solid wood simple and comprehensible and enable it to continue to compete
with other structural materials in the market place.

The whole of Europe is now becoming aware that a scientific system of stress
grading, visual or mechanical, is essential to the future marketing of building
timber. The choice is between visual or mechanical grading.

Unfortunately visual grading is limited to an examination of growth features
such as the size and position of knots and the number of annual rings per inch.
It is not possible to estimate the density, while the measurement of the slope
of grain involves a time consuming scratch test; yet these two factors are on the
whole more important in assessing the strength than the size of knots. So, while
visual grading can reduce the uncertainty in the strength of a piece of timber,
it cannot in my view reduce it much.

'Very much the same considerations of cost versus savings as applied to the
slow adoption of finger-jointing also apply to the development of the mechanical
strength grading of timber. Ten years ago there was a brief spurt of interest in
North America, which died down but has recently revived again. A separate line
of development in Australia led to the successful establishment of the Plessey
Computermatic timber stress grading machine there and, notably, in Europe, where

during the last two years 40 companies have decided on installations, large and small. The company of which I am Managing Director, Measuring and Process Control Ltd, may, as sole European agents for the Computermatic, claim some credit for this, but the basic reason is that mechanical grading gives the unusual combination of advantages of lower cost per cubic meter, better yields of higher-priced grades, greater reliability, and reduced costs for the builder.

The timber producer who is not deeply committed to previous methods of stress grading and associated processes can more easily incorporate this new technology in his mills and strike out with confidence along the new path.

Machine grading can be combined advantageously with some other necessary operation such as planing, end trimming or length sorting. This avoids extra handling and labour and so reduces any cost which is <u>additional</u> to the costs associated with the capital for the equipment and the <u>running costs</u> and maintenance. This additional cost is very low.

The Plessey Computermatic Mk P IVa, including a spare computer, costs about £20,000 (mid - 1974) installed. Together with the infeed and outfeed rigs and other mechanical aids which speed up production the line could be installed for £32,000.

Taking the annual costs at 25% of initial cost we can say

 Machine Costs only:

 Full line, annual cost   = $\dfrac{£8,000}{25,000}$ m^3 = 32p/m^3 or DMk. 1.9/m^3
 Annual production

 Grading machine alone,
 annual cost   = $\dfrac{£5,000}{10,000}$ m^3 = 50p/m^3 or DMk. 2.97/m^3
 Annual production

The labour costs will depend on what other operations are performed at the same time and how costs are allocated. An example will illustrate this.

 Estimated capital cost of end trimmer and machine grading with automatic
 infeed and package stacking attachments and end marking £50,000

Annual cost, say	£12,500
Building costs, say	£ 1,000
Labour - 4 men, say	£ 9,000
Overhead	£ 6,000
	£28,500
Production	25,000 m^3

Cost = £1.14/m^3 (6.77 DMk) for grading, end trimming, end marking and unit packaging.

Although a single Computermatic can produce 25,000 m^3/yr of graded timber, this is by no means the economic minimum and machines have been installed where the expected production is as low as 5,000 to 7,000m^3/yr. Another way of using the machine is for a number of sawmills to form a terminal where grading, length sorting and packaging can be carried out co-operatively.

The higher prices obtainable today for properly market machine graded timber in the UK are:

                                            <u>Yield</u>

M75 Grade £9/m^3 (DMk.53.5) extra   usually 75% at least
M50 Grade £6/m^3 (DMk.35.6) extra   usually 20%
Rejects up to £5/m^3 (DMk.29.7) less   usually under 5%
Average <u>added</u> price = £8.65/m^3 (DMk.51.4)

From the user's point of view, it seems that already an economy of about
18% can be made in the amount of building timber required for low-rise housing
while specifiers, those responsible for the safety of buildings, are not slow
to take advantage of the increased certainty of the material.

The degree to which producers have begun to take advantage of this improvement in marketability, utility and price of machine graded quality assured timber varies greatly, however, in different countries. In North America, mechanical grading still faces a barrier in the shape of the firmly established and
highly sophisticated visual grading rules and their supervisory structure. In
Europe, where each country made its own rules, only recently has a size standard
been attained and now there is a growing momentum towards the establishment of
standardised <u>grading</u> rules. The existence of an objective scientific method of
<u>mechanical</u> grading gives Europe the chance to overtake the Americans in maxi-
mising the utility of structural timber.

The establishment of a uniform European system of grading and quality control is likely to be rapid. The Timber Committee of the ECE (Economic Commission of Europe) has appointed a drafting committee which has put forward a draft
based on the new British Standard 4978, and there seem to be distinct hopes that
agreement, at least in principle, may be reached by the end of the year on two
visual grades with mechanical properties which are an improvement on those in
the new British Standard. As part of this process, suggestions for rational-
ising the bewildering variety of existing grades in European countries are now
being canvassed. The principle of mechanical grading has been accepted by the
Committee and provision for this improved method will be included. (Such a proposal was agreed by the ECE Timber Committee on 15 October 1974, setting up an
international standard for an 18-month trial period — Ed.)

In addition to defining grades which would be acceptable throughout Europe,
a common standard would need to specify a range of preferred sizes of timber. In
Britain a start has been made by the recently formed Timber Stress Grading Association which has announced the 18 sizes of stress graded timber that its members are making readily available in two constructional grades and which can be
safely specified now.

The future

In terms of practical application, progress is likely, as always, to be
patchy. But it is important to keep before us a vision of what we should like
to see in the timber processing plant of the future.

We can foresee, for example, a plant combining the
jointing and mechanical stress grading to provide, automatically, the grades
and lengths of timber required. We can bear in mind also that computerised

optical, as well as mechanical, quality control is now beginning to be possible.

Such a plant would accept parcels of upgraded timber including pieces of widely differing strength, appearance and size. It would be controlled by quite a simple computer with a memory which would be constantly updated according to the demands of the market for timber of specified size and quality and the selling prices of the different grades.

The plant would automatically grade pieces of timber and select those meeting the market requirements. The remainder would be passed to the finger-jointing operation where the joined lengths could be cross-cut so as either to optimise the cash yield from a consideration of the required grades and lengths and the prices of the various grades, or to meet the actual order situation at that time. While these two broad strategies - optimising cash yield from the available timber or meeting the demand in the most economical way - could not be operated simultaneously, it is possible to change from one to the other at some point. For instance, once orders have been met, grading, jointing, cross-cutting and regrading could continue on the "cash yield" principle or, again, on a pattern of expected demand, worked out by the computer or the operator.

The output figures from the grading/separating plant and the jointing/cross-cutting operation would be continually fed back to the computer, modifying the memory and ensuring "real time" operation. The computer could provide any form of standard and special report required on plant input and output, profit, stock position, market demand and the extent to which it was being met, while override facilities could be provided for additional control of the plant.

Such a plant can be a practical proposition much sooner than we may think possible. Certainly progress is fast. Computer-optimised cross-cutting <u>without</u> quality control sensing is already available (Wadkin Hurn, Leicester, England). It is claimed that the economic gain from reducing cross-cutting waste is greater than that obtainable from finger-jointing.

Nothing suggested here is beyond the bounds of present technology. Whether the goal is actually achieved in any particular place or time will depend on market factors which vary greatly from year to year and from country to country; but the overriding pressures in Europe tend inexorably towards this kind of long-term solution.

Discussion

QUESTION: The presentation was interesting, and it put forward entirely different possibilities. But the proposal also seems to pose serious difficulties. I can only start from existing conditions in the Federal Republic (of Germany), where the DIN standards apply. For stress resistance of construction timber we have only two grades - that is Class I and II. Certain requirements have been formulated here that can easily be determined, although they are not measured in the same way as described in Mr Serry's paper. But if this new technique (stress-grading) is to become accepted here, then it is of utmost importance that discussion is opened through the industry associations and other agencies to try to establish a wider spectrum of quality; that is, to have not two grades or classes for construction timber, but more, and this would require revision of existing regulations.

The question would then be, what the grade must be as far as mechanical properties are concerned. Until this is done, acceptance of these new techniques (stress-grading) will be a long way off.

I would like to ask Herr Fronius whether he can establish the current situations regarding this in other countries and what possibilities there are for more rapid progress in this in other countries.

SERRY: I have been to Germany quite a number of times. I have spoken at the Otto Graff Institute (in Stuttgart) with the authorities there, and I've spoken with Herr Fronius, and at other gatherings. It certainly is true that the first step towards use of a new grading system is to get a new standard.

I have had this problem in every country. In the United Kingdom the problem is solved; we have a new standard. In Germany I am hoping that some interim arrangements can be made to allow the technique to get a start. An appeal I would make to those in authority in the Federal Republic is to permit machine grading to be used to produce timber at least to the existing DIN norm. That has been done in a number of countries. I am expecting the results from Sweden very soon. I know that results will come this summer from Finland, and arrangements have been made in France for this to be done.

Since this system does not fit the existing practices of the industry, they have to be changed to fit the system. The need for this in Germany is considerable, and I only hope that perhaps this conference may be one of the steps towards reaching that point.

However, that does not necessarily prevent the start of the practice here, because I believe that in Germany arrangements can be made for an engineer to be made responsible for his own work. With the stress-grading concept an engineer can be very sure of what he is doing, and so in things like the trussed rafter industry I think a start can be made right away.

I do not claim to be an expert on this, and I would like to hear Herr Fronius's views on this same subject.

HERR FRONIUS, Programme Director: The Rosenheim Institute for Wood Processing and the Otto Graff Institute in Stuttgart are, in fact, the first to seek progress in this, jointly, and also to expose both professionals and the general public to the concepts. But as soon as we get this under way, I think the authorities will not be opposed to the idea. It may take a few years yet, but I am quite certain that machine grading will make a breakthrough.

As far as the situation in other countries is concerned I would like to ask Professor Thunell from Sweden, where, if I am correct, machine stress grading, or at least stress grading, has already been used.

THUNELL (Professor Bertil, Swedish Forest Products Research Laboratory, Stockholm, Sweden): In Sweden, we have been working on this for several years, with the institutes, and have carried out a number of trials. These are concerned not only with machine grading, or giving a basis to machine grading, but also with the new British visual grades, GS and SS classes, sorting grades.

As Mr Serry has already said, it is likely that during this summer two machine grades will be used in production, at least in a pilot project way. This will involve introduction of, first, the British grades MGS and MSS, and, secondly, the corresponding grades for what we call T300 and T200, which are the Swedish stress grades.

Regarding pre-fabricated house construction and the application of stress grading, we will be able to begin introduction of the grades I have referred to. But what we particularly want to see is use of stress grading machines extended to the lower qualities of sawntimber, and this is now being tried out.

QUESTION: What have been some of the attitudes in the timber trade regarding introduction of the new grades?

SERRY: If I understand the question correctly, it is: what is the acceptance by the timber trade in Britain of this new concept? I'd say, controversial, uproarious!

To some extent the controversy of the past three or four years is now dying down, but adoption of the grades has been a big disturbance to the way of trading. Demand in the UK, now that consumers know the problems have been solved, is considerable and sufficient to cause almost all timber suppliers to consider either visual or mechanical grading. Suppliers that have pioneered the process are well satisfied with what they have done and those who invested early are now getting knowledge and understanding of the technique and are happy with the position. Those who decided that the concept would not be accepted and would disappear are rather unhappy. That is the present situation in the United Kingdom.

In the United States, the position is that there is not enough mechanically-graded wood available for the demand, because quite gradually, with no one noticing, there was an increase in the utilization of wood because of the economy and in a large number of well-organised mills it has been difficult to find a way to fit the process in.

In the States the subject of machine grading is highly controversial, too. The Forest Products Research Laboratory at Madison, Wisconsin, is very much in favour of the process. I have heard of strong opposition in the Southern states to Mr. Bill Galligan, who has pioneered it there.

KES, E., Council for Scientific & Industrial Research, Rep. of South Africa: I would like to add that we in South Africa have also developed a system for stress grading that has a slightly different approach, especially with regard to machine size. Our machine is a static stress-grading machine which costs only a fraction of the considerable investment involved in your machine. This machine accomplishes both mechanical stress grading of timber and also quality control of fingerjoint. You mentioned as a thing of the future that stress-graded material could be finger-jointed to give long lengths of a specific stress grade. In South Africa we are doing this already. We are using this method to make laminated beams of a specific strength, and this method has already been adopted in standard specifications in South Africa.

SERRY: I am delighted to hear that my prophecy has been fulfilled. I have
heard about the South African machines, and I believe there has been one in
England. While any work in this field is obviously very welcome, I do not think
that the manufacturers would claim that the productivity of this machine will
compare with the much more complex stressing machine. However, it is an
excellent machine, I believe, and it works on the principle - as I was told - of
visually selecting what appears to be the worst place and in testing that, which
is quite a sensible principle. However, this does definitely seem the machine
for rather smaller scale production than we are thinking of with the machine
capable of working at 157 m/min. Is this in fact the case?

KES: I fully agree that productivity is very much lower, but on the other hand
one must keep in mind that stress-grading is a strange and new process. We
thought that to invest perhaps DMk 5,000 instead of DMk 100,000 might be a
better and an easier solution for a sawmiller. The smaller investment is easier
to make for trying the process out and then, later on, as a second step he can
go in for the more expensive machine. For the development of markets and
obtaining market response we thought the lower investment could work well, and
we have some encouraging results in South Africa.(Applause).

SERRY: I certainly cannot find any flaw in your argument and certainly I would
welcome that people should make a fairly cost-free experiment to find out the
virtue of the system. I will sit and wait with great patience until you begin
to want to do it on a larger scale.

BRAUN, Monsieur Robert, S.A. Ferdinand Braun, Niederhaslach, France : I would
like to add something as a point of information in response to Herr Fronius'
request for reactions in the timber trade to the new grades. Since I am also a
delegate for the sawmilling section of the Centre Technique du Bois, I could
also comment on views in Germany on practical grading matters and the new
systems of testing and grading timber.

It seems to be felt that stress grading might offer a great opportunity
for timber as a material because architects are only too often reluctant to use
it through lack of interest or lack of data regarding stress resistance of
wood. So machine-grading is looked on as offering a better solution resulting
in the drawing up of standard specifications and allowing architects to better
utilize sawngoods.

This is the general reaction. The only difficulty, then, is the question
of the price. This new technique must first be publicized and promoted so that
people are made aware of it and that those using it will be ready to pay the
slightly higher price that will be required. But already in France there are
certain sectors that have shown interest, in particular engineers involved in
the design of underground mines of a highly mechanized sort. These enterprises
are definitely interested in purchasing machine graded sawngoods.

SERRY: The question is highly interesting. I have some figures on it I would
like to show you. Considerably higher prices are obtained for machine-graded
wood in Britain. Because of timber being so variable, I can only tell you how
much additional revenue is received. These figures give the position in the
U.K. today, the only European country where stress graded timber is being sold.
The grade M 75, the highest grade, would be applicable to more than 75% of
normal European fifth quality. The additional cash yield on this grade is £9/m^3
say DMk 54/m^3 additional. The second grade, M50, which is something less than 20%
so powerful, would bet £6/m^3 additional, or DMk 36/m^3 more.

11

Optimization of recovery rates using modern thin-kerf sawing systems

By Lutz Claassen, B.Sc., Research Department, Canadian Car (Pacific), Vancouver, British Columbia, Canada.*

Thin kerf sawing is certainly no longer a hidden concept. Many mills have already installed systems and yet others are actively engaged in planning and building newer and improved systems. As one recent news article puts it ".....
... announces a new mill will be built using linear thin kerf sawing." These words describe a sawmill system that collectively improves production and yield. It is often the case that in order to improve a subcomponent, the system as a whole must be improved to realize the benefits.

The subcomponent in this presentation is the 'Thin Kerf Saw'; the system includes accurate scaling, optimum machine adjustment, and true control of the workpiece as it is fed through the machine. This presentation will touch upon only the thin kerf saw, but, of course, the system must remain in perspective.

Average fibre yields for a Chip-N-Saw

Work progressed very rapidly in the last two years in improving the system. Optical scanners and computer scaling programs are now being used in many mills to control the machine adjustment. Thus mill management is able to accurately evaluate the cutting patterns against maximum lumber yield or maximum value yield. For the Chip-N-Saw a calculated typical fibre yield, on a per log basis, assuming maximum volume recovery, is as shown in Figure 1, overleaf.

*This presentation, in a modified form, was made by Mr. Ole K. Marcussen, Ing. MAI, General Manager, Chip-N-Saw A/S, Nästved, Denmark.

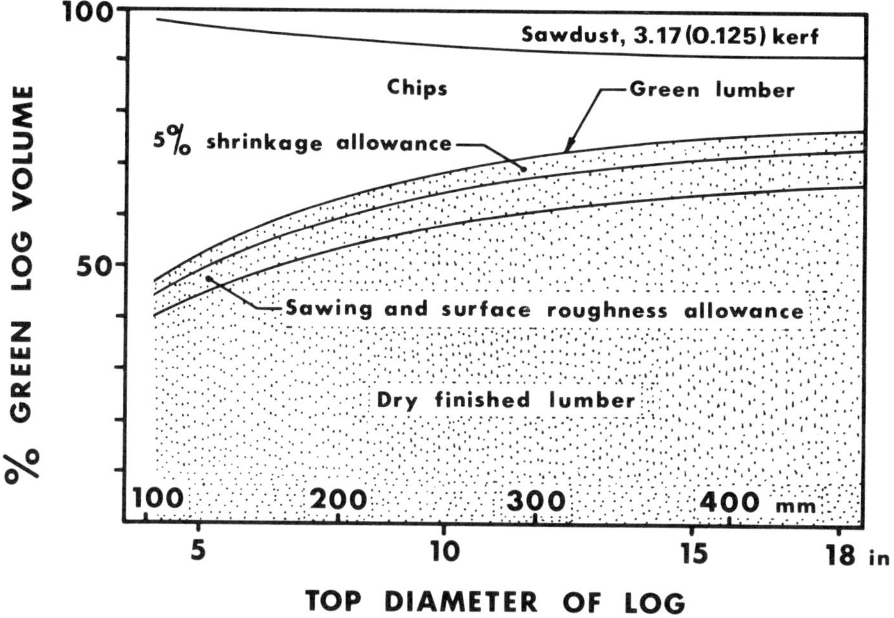

This data can, of course, be only typical since conditions vary considerably from one mill to the next. The varying conditions may be: log supply and quality, different markets, and lastly, sophistication of the system. Some mills cutting lumber for the North American housing market are able to exceed these yields, since a certain amount of wane is permissible. On a per-log basis these yields have been substantiated; however, for a day's production run many imperfect logs must be processed, and these require system sophistication.

Perhaps the item sometimes overlooked in this rapid change to greater sophistication is the fact that the lumber issuing from a machine is not of the quantity and quality that had been envisaged by the computer before the log entered the machine. What went wrong?

The log or cant did not feed through the machine properly, or the saws are not cutting with a consistent accuracy.

Sawing parameters

The ability to cut accurately at the mill production feed speed is the only and foremost requirement of the sawing process.

Figure 2 - Specifications of the
       circular saw systems described

	SAW 1	SAW 2	SAW 3	SAW 4
Saw type	Swaged tooth	Carbide tooth	Carbide tooth	Carbide tooth
Saw dia, mm inches	635 25	660 26	635 25	508 20
Saw plate, mm inches	2.10 0,083	2.10 0.083	2.10 0.083	2.10 0.083
Number of teeth	36	26	36	26
Saw speed, rpm	1,770	1,770	1,770	2,150
Collar dia, mm inches	241 9.5	241 9.5	368 14.5	140 5.5
Kerf, mm inches	3.17 0.125	3.17 0.125	3.17 0.125	3.17 0.125
Feed speed, m/min. feet/min.	160 48.8	155 47.2	150 45.7	160 48.8

Figure 3 - Specifications of the
       bandmills described

	Bandmill 1	Bandmill 2
Band speed, m/sec (feet/min)	45.5 (8,950)	41.5 (8,160)
Band width, mm (inches)	254 (10)	254 (10)
Band thickness, mm (inches)	1.47 (0.058)	1.65 (0.065)
Gullet area, mm^2 ($inches^2$)	580 (0.90)	458 (0.71)
Tooth pitch, mm (inches)	50.8 (2)	44.4 (1.75)
Distance between guides, mm (inches)	775 (30.5)	508 (20)
Kerf, mm (inches)	3.17 (0.125)	3.17 (0.125)
Feed speeds, m/min, (feet/min)	30.5, 79.8 (100,262)	39.6, 85.3 (130,280)
Total band strain, kg (lb)	6,350 (14,000)	5,440 (12,000)

In evaluating the process the objectives are, therefore, to minimize the arithmetic sum of:

The actual kerf and the allowance for sawing inaccuracy
and the allowance for surface roughness

These parameters relate to the end product and solely determine the adequacy of the process. To achieve a certain level of performance our research appears to indicate that only three primary parameters are required:

Saw loading, saw plate stiffness, and saw plate stability
margin

Saw loading determines the forces acting on the plate, whereas saw plate stiffness and stability margin determine the resistance the saw plate offers, to lateral loads induced by cutting, or to machine vibrations. These sawing parameters will be discussed with respect to four circular saw systems; Figure 2 presents some specifications. Figure 3 presents specifications of two bandmills that will be discussed.

Actual kerf width

In the case of thin kerf saws that are working satisfactorily the actual kerf can be assumed to be equal to the tooth width. Field and laboratory data on average board thickness support this contention. Theoretically this may be justified in the fact that the thin kerf saw has a resonance spectrum (to be discussed later) consisting of very low frequencies. These frequencies create undulating lumber and not a widening of the kerf.

The average board thickness measured is generally from 0.127 mm* (0.005) to 0.305 mm (0.012) thicker than the space between the saw teeth, see Figure 4 overleaf. This increase in board thickness is probably a result of wood fibre bending and resultant spring back as it is being cut by the tooth.

Allowance for sawing inaccuracy

This item must be evaluated as a statistical quantity. The quantity most relevant is the standard deviation of the board thickness, shown in Figure 5

To calculate standard board thickness deviation a minimum of five thickness readings should be recorded of one edge of a board and an equal number of boards must be sampled. The calculation can then be arranged in such a way that we calculate the standard deviation of each individual board. Then averaging the standard deviations over all the boards we would obtain a quantity which we will call pooled standard deviation within boards (   ). This quantity tells us whether the saws are cutting straight.

*All dimensions are in millimeters and followed in brackets in inches, unless
 indicated otherwise.

AVERAGE BOARD THICKNESS

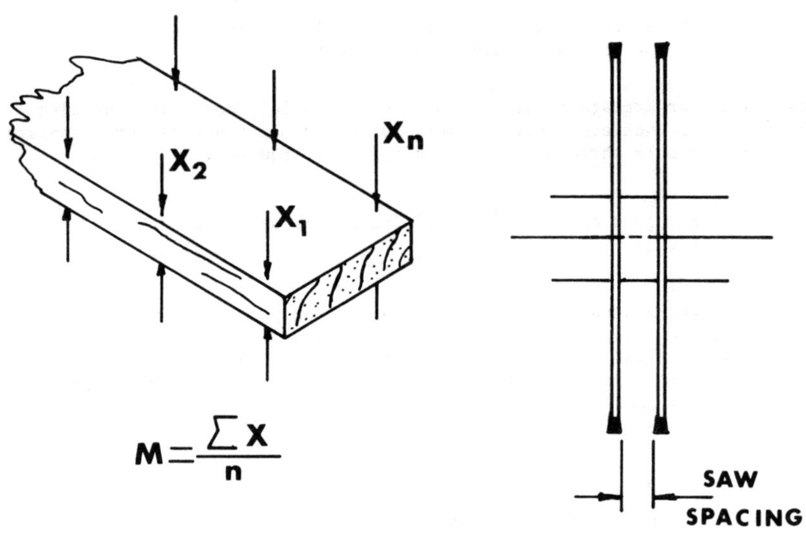

SAMPLE DATA		
	Saw Spacing	Average Thickness, M
Saw 1, mm (inches)	42.85 (1.687)	43.31 (1.705)
Saw 2, mm (inches)	23.80 (0.937)	24.10 (0.949)
Saw 3, mm (inches)	42.85 (1.687)	43.05 (1.695)
Saw 4, mm (inches)	37.54 (1.478)	37.92 (1.493)

Figure 4 - Average board thickness compared with saw spacing

Figure 5 - Definitions of the terms used in analyzing
board thickness accuracy

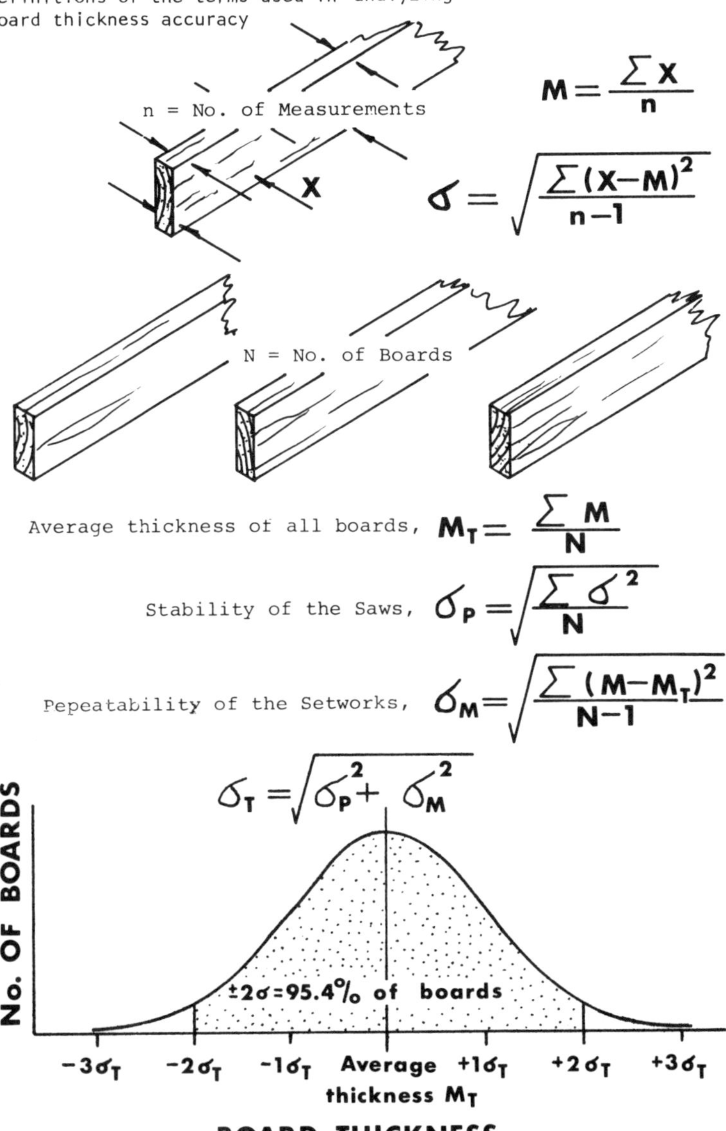

Furthermore, we can take the mean thickness of each board and calculate the standard deviation of the mean thickness, which we will call the standard deviation between boards (σ_M). This quantity tells us whether the setworks are adjusting correctly, or in general whether the cant or log feed system is working well.

This data can then be used to analyze the cutting process. The two quantities combined give us the total standard deviation of the process (σ_T). This quantity should be used in the computer process calculation to set the cutting patterns and analyze yields. From a probability point of view 95% of the production would have board thicknesses equal to the average plus or minus two standard deviations. (Average board thickness $\pm$ 2 σ_T).

Figures 6 and 7 show data that is currently being achieved with circular saws, 3.175-mm (1/8 inch) kerf, cutting 10.16-cm (4 inches) and 15.24-cm (6 inches) depths and bandmills, 3.175-mm (1/8 inch) kerf, cutting various depths.

Figure 6 - Circular sawing accuracy data
measured in several sawmills

	(Saw stability) σ_P	(Setworks repeatability) σ_M	(Total) σ_T
Depth of Cut = 97 (3.820)			
Saw 1 (Fir)*	0.330 (0.013)	0.483 (0.019)	0.584 (0.023)
Saw 2 (Pine)*	0.432 (0.017)	0.483 (0.019)	0.647 (0.025)
(New Saws, New Guides)	0.051 (0.002)	0.178 (0.007)	0.185 (0.007)
Saw 3 (Fir)*	0.432 (0.017)	0.305 (0.012)	0.528 (0.021)
Depth of Cut = 150 (5.925)			
Saw 2 (Pine)* (20 Boards, newly	0.914 (0.036)	0.838 (0.033)	1.240 (0.048)
sharpened saws, Fir)	0.584 (0.023)	0.736 (0.029)	0.940 (0.037)
(New Saws, New Guides) Saw 4 (Hemlock) (Experimental Test	0.178 (0.007)	0.279 (0.011)	0.331 (0.013)
Bench, 20 cuts)	0.191 (0.0075)	0.152 (0.006)	0.244 (0.0096)

* Data recorded during actual mill production over a period of 16 hours, no saws were benched or replaced during the test period.

If a satisfactory requirement is that 97.5% of a mill's production meets
the size requirements, then the allowance for sawing inaccuracy should equal
$\pm\ 2\ \sigma_T$. The data of Figure 1 includes an allowance of the magnitude of Figures
6 and 7.

Figure 7 - Bandmill accuracy data
          measured in two sawmills

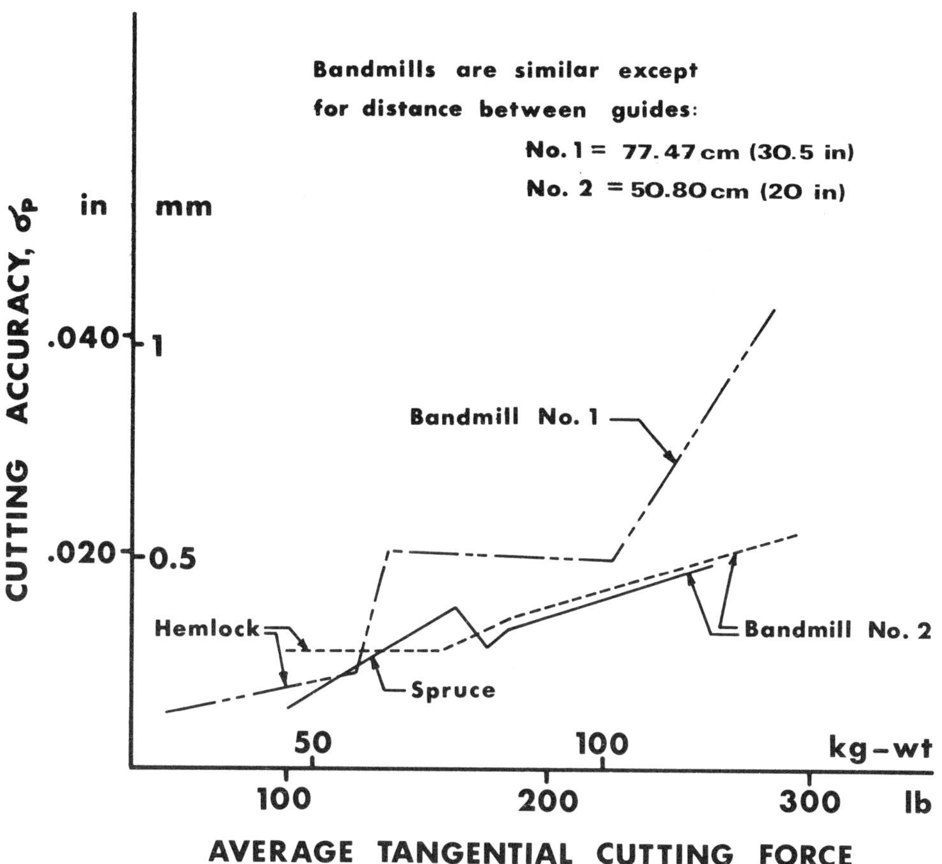

Surface roughness

Surface roughness appears to be markedly less in the thin kerf machines than in the heavier saw machines. This fact is related to the stability spectra of the plates, and saw plate maintenance. With heavy, stiff saws, it is extremely crucial that the plate is levelled accurately and that it runs true on its arbor.

Thin saws on the other hand produce no noticeable roughness as a result of static run-out. As long as they are mounted on collars running reasonably true or are floating on an arbour, the workpiece is practically void of all tooth marks.

There is, however, one exception. Very noticeable tooth marks occur when the saw is being rotated at speeds at which rotational speed matches a resonant frequency of the plate. This is a result of lack of balance within the machine. As a certain amount of imbalance will have to be tolerated in present day saw-mill machinery the rotational speed of the thin saw must be carefully chosen.

Saw loading

Saw loading appears to be a very important variable, as it might be postulated that a certain portion of the cutting force (tangential force) acting on the saw blade reflects itself in an unbalanced side force. This side force could be caused by a tooth defect, by wood grain direction, or by improper machine feeding.

Figure 7 presents data, measured during production at two sawmills, of saw loading versus sawing accuracy. It is this type of data that we are presently correlating against saw plate stiffness so as to arrive at a most favourably loaded sawing system.

The variables that must be considered in minimizing the cutting force are:

> Feed speed, kerf, depth of cut, bite and saw
> peripheral velocity.

From our tests, and also cross plots of tests by others, it appears that we can confidently use test data such as that of Figure 8 to calculate the cutting force. The calculation would not be valid over the extremes of bite and peripheral velocities, but can be applied over the ranges of these variables that are of general interest. By measuring power consumption for a number of different feed speeds, data of this type can be generated for the particular sawing system and wood species, and then used to evaluate the cutting process.

This data shows that a large bite and a high peripheral velocity produce the smallest cutting force. However, gullet capacity and the force per tooth are limits on the bite, whereas machine balance and saw stability criteria impose limits on the peripheral velocity.

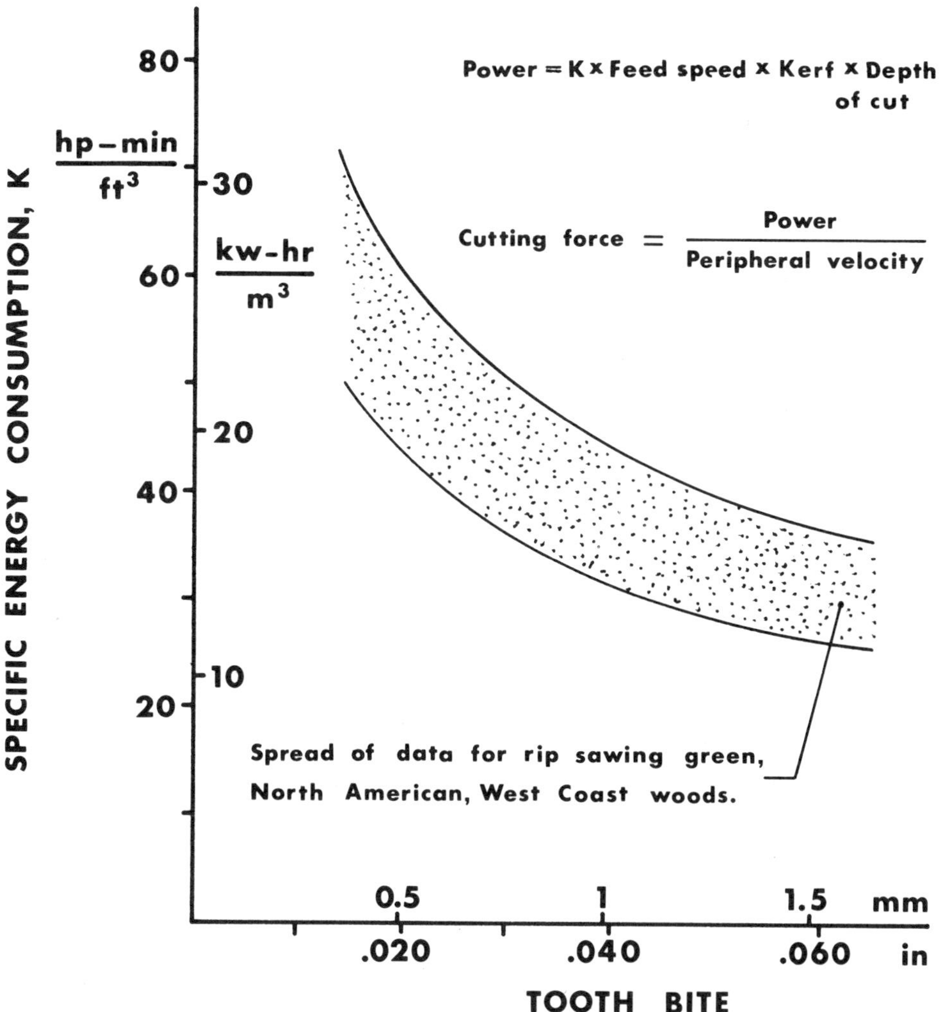

Fig. 8 - Specific energy data measured in several sawmills

Saw plate stiffness and saw plate stability margin

All circular saws completely lose their stiffness at certain rim speeds.
The approximate rim speeds at which this takes place for saws that contain no
internal tension are given in Figure 9

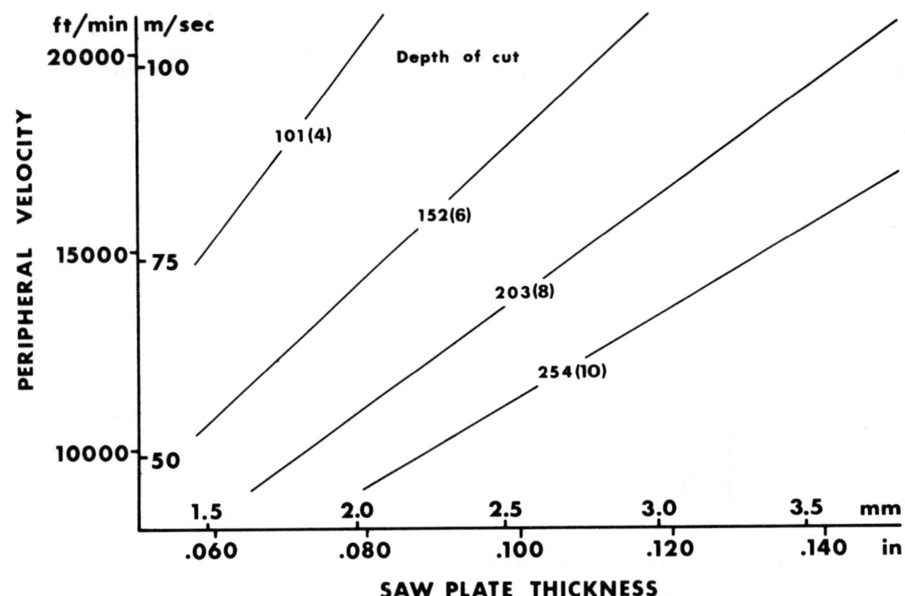

Fig. 9 - Minimum values for the critical speeds of saws
containing no residual stresses and sized to cut
the depths as shown.

The phenomenon is a standing wave resonance instability, as in Figure 10.
When a saw is tapped or forced to vibrate at different frequencies, certain frequencies cause the saw to vibrate excessively. At these frequencies the saw is
resonating in different modes of vibration, and distorts into wave shapes as
given in Figure 10. These shapes can be easily observed at rim speeds slightly
below those of Figure 9, when they appear as waves travelling in an opposite
direction to the rotation of the saw.

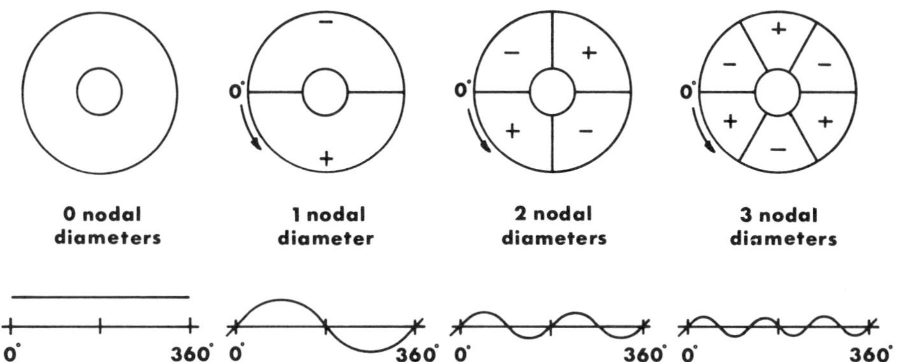

APPEARANCE OF THE RIM OF THE SAW BLADE

Fig. 10 - Basic shapes into which a
circular saw distorts when
rotated at the critical speeds.

The data of Figures 11 to 14 (see overleaf) provide partial vibration
spectra of three saws. The descending and ascending lines describe the reverse
and forward rotating natural frequencies respectively. Standing wave resonance
instabilities occur where the lines cross the horizontal axis. The data of
Figure 9 is a plot of the lowest crossing, this being called the critical speed
of the saw.

Large amplitude vibrations will occur whenever a saw is subjected to vibration at these frequencies. A vibration source may be an unbalanced saw arbor.
In this case, large vibrations would occur at the saw speeds at which the rotational frequency line crosses any of the resonant frequency lines.

Now, saw tensioning and saw heating change the stability graph. Saw tensioning shifts the modes 2 and higher curves vertically up (the curves remaining
parallel to the untensioned saw data), as in Figure 13. Tensioning, therefore,
raises the critical speeds and makes it possible to rotate the saw at a higher
speed. Rim heating does the opposite and may thus cause the saw to become un-
stable.

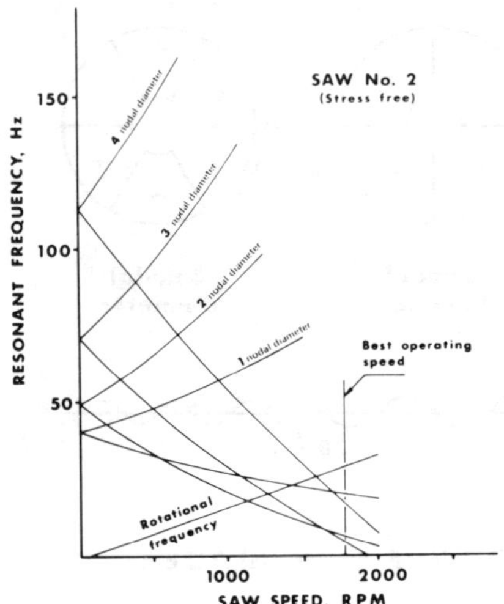

Fig. 11 - Resonant frequencies of saw No. 2.

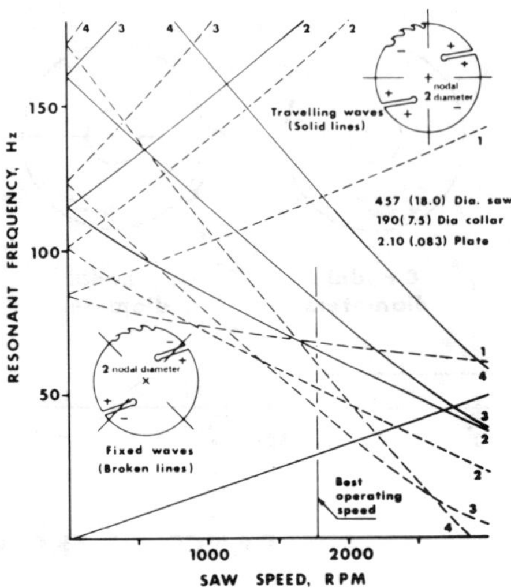

Fig. 14 - Resonant frequencies of
a slotted saw.

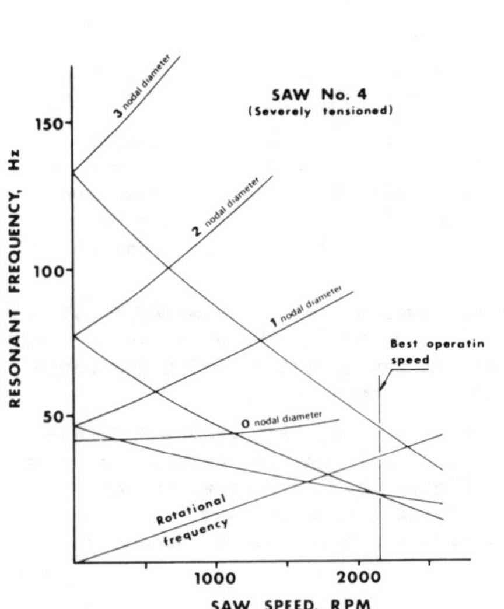

Fig. 12 - Resonant frequencies of saw No. 4,
no residual stresses.

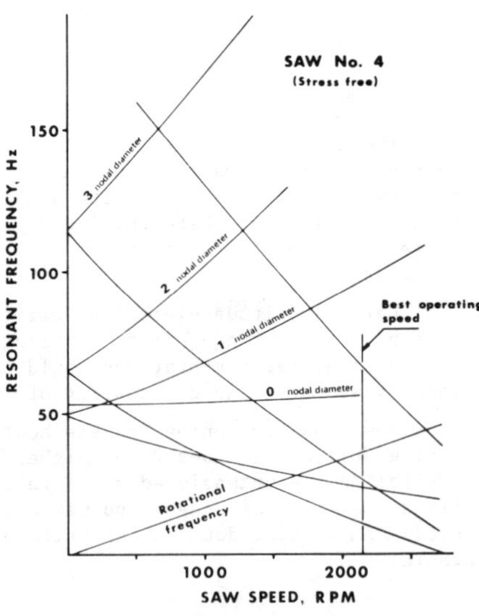

Fig. 13 - Resonant frequencies of saw No. 4,
severe treatment of tensioning.

Tensioning, on the other hand, lowers the 0 and 1 diametral modes of vibration. These modes describe the deflection of a saw when acted upon by a single static force, provided the saw is not rotating at a critical speed. Hence tensioning reduces the saw stiffness, although it makes running possible at a higher peripheral velocity. In general, over-tensioning is just as detrimental to the cutting accuracy.

Therefore, to briefly summarize, for accurate cutting the saw must be dynamically stable and offer a good resistance to a lateral force. The peripheral velocity should be as high as possible. How might these requirements be achieved?

Referring to Figure 11, a best operating speed would be 1,770 rpm. At this speed, the rotational frequency does not match any of the resonant frequencies. As this speed is below the critical speed of the saw, this plate would perform best with only a small amount of induced tension in the blade, only enough to ensure that the stresses in the rim of the saw are not of the compressing type.

Figure 14 presents a stability spectrum of a slotted saw. This particular saw, with two slots, has gained wide acceptance because of its relative maintenance-free operation (high tolerance to rim heating). Because of the density of the resonant frequency lines, successful operation appears possible only below the lowest inter-section of a resonant frequency line with the rotational frequency. Because of this fact, the slotted saw must be made of a heavier plate in comparison to the solid plate saw. The higher rim temperature tolerance of the slotted saw is being matched by solid plate saws through tensioning and rim cooling.

Having determined the stability characteristics of the blade, we now require that the blade be as stiff as possible (without effecting an increase in the plate thickness or a decrease in the depth of cut). Different collar and guide arrangement can be very helpful. Guides, of course, can only be used on solid plate saws.

Collars and guides

Collars and guides may be evaluated theoretically in terms of their effect on the stiffness and stability of the saw blades.

For a 152-mm (6-inch) depth of cut and a kerf of 3.17 mm (0.125 inches) a saw with a collar fixed rigidly on an arbor can be deflected at the rim, when not rotating, approximately 0.39 mm/kg (0.007 in/lb), at the midpoint of the cut. The same saw, but with a floating collar and 2 or 3 point guides, would deflect 0.76 mm/kg (0.030 in/lb). Therefore, the fixed saw is, no doubt, superior. The guide nearest the entry of the saw blade into the work piece, however, causes a markedly large increase in the saw plate stiffness at extreme saw deflections, such as being caused by improper work piece feeding. Thus it may be reasoned that a guide and a rigidly mounted saw form the best sawing configurations. Machine alignment is intensely more critical; this, however, is a problem that is being overcome.

What are some other reasons that create a good case for guides? To ensure consistent operation of the saw, the stability spectrum must not shift up or

down, particularly down. A water cooling film on the saw appears to be very effective. Guides offer a means to effectively brush a very thin film of coolant
onto the blades.

In situations when extreme forces cause the saw to move laterally, the
guides cause the plate to heat at the midsection. This brings about thermal plate
tensioning, thereby maintaining or increasing the stability margin.

What about no collars? Saws with no collars, splined on an arbor, display
a somewhat similar frequency spectrum to the other saw configurations when the
kerf and depth of cut are the same, but the saws are considerably smaller in diameter. Saw stiffness is considerably less, unless extremely large guides are
used.

In conclusion, it should be stressed that manipulations of the sawing parameters lead to thin saw systems that cut as accurately as the saws of yesterday
with the same demanding depths of cut and high feed speeds.

Furthermore, optimizations of this type lead to comparisons and a search
for something better.

In our research laboratory we have tested a patented guide system that is
providing somewhat higher saw stiffness to all the other configurations we have
tested and a considerable amount of dynamic dampening.

The guide operates with the principle of the spiral thrust bearing. A small
amount of liquid lubricant, though, is added to aid in the lubrication. Figure 15
(see view 22 on page ix at the back of this book) presents a photograph of this
guide.

Laboratory sawing tests with this guide have demonstrated accuracy and surface roughness to be of sufficient quality to permit direct marketing of the sawn
product, or post-finishing with abrasive sanding. When field operation has proven
these results, a major step will have been signaled in the quest for greater optimization of recovery.

Discussion

: I would like to know the saw diameters used, the blade type, and kerf.

N: The system is not one that operates with one specific diameter, and
the same thickness one can use different diameters. Depth of cut is
rmining factor. Most of our machines use a 600-mm diameter blade, and
a depth of cut of 17 cm. Depth of cut multiplied by a factor of three
ve an idea of the diameter required. Saw blade thickness, as I pointed
2.5 mm with a cutting kerf of 3.1 mm. This is possible with circular
n with feed rates as high as 50 m/min.

12

Waste utilization: getting optimum value

By Herr Dipl.Forstwirt Karl-Reinhard Volz, Forest Officer, Forestry Administration, State of Baden-Würtemberg, Fed. Rep. Germany.*

Bark and sawdust residues from sawmills can either be burnt to supply steam for power generation or they can be used in manufacture of various boards. For bark there is a third possibility in its use in soil improvement in agriculture. All other uses are either confined to certain regions or are not practiced, due to lack of demand.

Dumping of these residues in official or unofficial refuse tips, which is still widely practiced, is not an acceptable long-term solution for either type of residue. Not to be overlooked is the fact that for a large number of sawmills no economic or relatively risk-free alternative to dumping exists at present. Nevertheless, anti-pollution legislation requirements (bark is regarded as an industrial waste in the German Federal Republic) and the cost of disposal, which has risen steeply recently, make the need for increased and improved utilization of bark and sawdust a pressing one.

The problem of achieving increased utilization is limited in the case of sawdust, its characteristics being well enough known. By contrast, knowledge of the properties of bark that are significant to its possible utilization are very limited. Further, the considerable differences between the bark of different species, in anatomical, chemical, and physical properties, often make impossible any general application of knowledge that does exist.

*Herr Volz carried out his research at the Wilhelm Klauditz Institute for Wood Research, Braunschweig, FRG, under Professor H. Schultz, formerly Director of the Institute, who assisted him in his presentation. Professor Schultz became Director of the Institute for Wood Research and Technology, Munich, FRG, in early 1974.

Our investigations of the density and swelling characteristics of a few different types of bark clearly demonstrated the large differences both between various barks and between the bark and the wood of the same species. The density of spruce bark, for example, was about one third greater than that of the wood, and in the case of beech the bark was considerably denser as well. The degree of swelling showed even greater differences: for spruce, pine, and beech bark it was in some cases more than double the swelling of the wood of the tree concerned.

These differing characteristics add considerably to the difficulty of finding a suitable use for debarking refuse. So it is not surprising that the use of the bark as fuel is by far the commonest form of utilization practiced, although the calorific value of the bark, which is usually moist, is relatively low.

Special furnaces have been designed in which bark with moisture contents of sometimes more than 50% can be burned without difficulty. Technical solutions to the problems of clinkering of the grates through large quantities of dirt, the feeding of the small-sized material into the furnace, and the initially very difficult problem of dust emission have been found.

The question of the economics of burning mixtures of bark and sawdust on the other hand, cannot be completely and clearly answered. In general, the furnaces are very expensive and their operation involves such a large number of employees and therefore a high cost that burning simply as a means of disposal, without utilizing the energy produced, is usually an unprofitable procedure. Many calculations have been made of the heating value of bark from a financial standpoint and they have always shown that it is possible to use bark as a fuel for heating and drying at an economic cost. Due to the world-wide shortage of raw materials, the profitability of burning wood waste products can, under certain circumstances, still further improve.

These considerations are, however, subject to the reservation, which cannot be overlooked, that not every sawmill is planning the installation of a new drying plant for sawntimber and that, where heat is required, the necessary heating equipment, usually oil-fired, is already installed. It is possible to convert existing heating equipment but the cost is relatively high. In such cases the briquetting of the wood waste could be of interest. In general, the burning of bark and sawdust can be regarded as one of the possible forms of utilization, but it must be examined on its merits for each individual sawmill.

Processing bark and sawdust into chipboard would at first appear to be the obvious solution, since the technology of the process in any case requires fragmentation of the raw material. Sawdust is nowadays collected and paid for in bulk by dealers or directly by the chipboard manufacturers themselves. The growing shortage of roundwood has resulted - with regional differences - in an increased use of this material, despite its being basically not very popular for the manufacture of chipboard.

Use of sawdust in the middle layer of board is the most logical procedure because there the least inconvenience is caused by the very unsuitable shape of the particles. Their relatively large thickness results in a high proportion of small faces, which hardly contributes to the rigidity of the board. Inadequate

overlapping also has an adverse effect on rigidity. Furthermore, the high bulk
weight of sawdust reduces the degree to which the board can be compressed. Problems can also arise through separations occurring in the dispersal of the chip-
pings, and this can result in considerable fluctuations in board quality. The
total effect of all these difficulties, which we have merely indicated, has always led to reduced use of sawdust whenever there is an adequate supply of timber.

A more recent possibility for the use of sawdust in board manufacture has
arisen through the efforts made by the chipboard industry to provide still finer
surface layers, especially on boards intended for outer finish or decorative
purposes. For these micro- or very fine surface layers the raw material is fine-
ly pulverized in special mills. Sawdust is very effective here if it is clean
enough. With production of these extremely fine surface layers on the increase
utilization of sawdust will inevitably increase, if the problems of collection,
transport and storage can be satisfactorily solved.

The problems of using bark as a raw material for chipboard are considerably greater than for sawdust.

Experience to date is principally confined to processing wood that has not
been debarked, that is, with a proportion of bark in the boards of about 10% to
a maximum of 20% of the volume of timber used.

If, however, the present and future supply of bark is considered, especially outside the chipboard industry, it is clear that this level of use is
only a partial solution of the problem. The utilization of large quantities of
bark will only be achieved if boards are produced with a high proportion of
bark or entirely of bark.

The process of reducing the size of the bark depends largely on the condition in which it is delivered. In general, a primary reduction stage in a hog-
mill is necessary because the pieces of bark and wood are too large to be reduced to the final size in one operation. Today there is a special type of hog-
ger for this purpose in which the problem of feeding in the material is also
satisfactorily solved.

The final processing of the bark to particles similar to chips is a more
difficult matter. This similarity to chips means that the particles should be
of suitable length and as thin as possible, that is, they should have as high
a fineness ratio as possible to give good bonding between particles. Experience
shows, however, that bark particles are generally thick and short.

The shape and size of the particles will depend less on the type of reducing mill used than on the kind of bark, its moisture content, the time at
which debarking was carried out, and also on the conditions and period of storage.

In principle, bark particles can be used for the manufacture of hardboard,
and the properties of the board will differ according to the kind of bark employed. The bending strength of board made from spruce bark, for instance, is
better than those made from pine or beech bark, while, on the other hand, beech
bark boards can have very good transverse tensile strength. In general, boards
made with 100% bark attain the strength characteristics required in traditional

chipboard by the DIN standards only if density is increased above 800 kg/m^3
and if, in most cases, a larger quantity of glue is used. Urea phenolic resins
can be used without producing any essential modification of the characteristics, while practically no improvement of bark boards is effected by an alter-
ation of the pressing conditions.

An improvement of the quality of the boards can be made by mixing the bark
with other materials, particularly with sawdust. Use of bark for the middle layer of three-layer boards with sawdust in the outer layers have given good results, since board strength is given by the outer layers. Depending on the kind
of bark and the shape of the particles, the relevant DIN standards can be attained with a proportion of bark ranging from 30 to 80%, starting with the customary density of about 650 kg/m^3. The properties of bark boards, particularly
the bending strength, are also distinctly improved by laminating the surface
with wood, plastic, or metal.

A new type of board with a special construction which makes only few demands on the natural strength of the bark particles seems to offer good possibilities. In this process, when the glued particles are compressed they are at
the same time shaped by small conical projections on the pressure plates in
such a way that the cross-section of the board corresponds to that of a lattice
girder, but without its tension and compression flanges. As a substitute for
these flanges whose function is to resist bending stresses, both sides of the
board are faced with sheets of plywood. The strength characteristics of such
composite boards, particularly their flexural stiffness, surpass those of traditional chipboard of the same weight per unit area. The cavities thus formed
in the pressing process result in the formation of boards with a very low weight
per unit area. It seems probable that these boards could be used for wall sections or similar purposes. More details will be available after further investigations have been carried out.

Thus the use of bark in the manufacture of hardboard, which could only be
dealt with briefly here, seems on the whole to offer good prospects, although
results of technical and economic experience on a large scale are not yet available. There is also, above all, the question whether chipboard manufacturers
will be willing to install the additional machinery to process large quantities
of bark.

In considering this form of utilization, therefore, it would be worth-
while investigating the possibility of a number of board manufacturers combining
together to carry out processing of bark, together with sawdust, in a central
installation, provided that a sufficient supply of raw material is assured.

The employment of bark as a means of improving the soil is absolutely in
accordance with its properties and, through the return of the bark to the natural environment, appears to be a very sensible form of utilization.

No variety of bark contains so much nutrient that it can be employed as
manure. But, by reason of its specific physical action, bark, similarly to peat,
can be used without prior composting either for mixing with the soil or as a decorative layer on top of the soil.

A big advantage of both these forms of utilization lies in the low cost of
preparation. All that is absolutely necessary is a reduction in size. Suitable

machines for this purpose are supplied by various manufacturers. The required
size of the particles, and thus the choice of a suitable crushing or pulverizing
process, depends on the kind of bark and the uses to which it will be put.

Employed as a decorative top layer on the soil, bark improves the quality
of the soil by acting as a protection against evaporation, by maintaining the
structure of the soil and by effectively compensating for differences in temperature, besides strongly checking the growth of weeds. The result of all these
factors is that soil thus treated with bark need hardly be cultivated for a
whole year.

The second way of using bark for soil improvement is to mix the pulverized
bark with the soil, whereby the soil's water retention and cation exchange capacity is improved and a greater loosening and aeration of the soil is achieved,
particularly in heavy soils.

The big problem with the use of bark on and in the soil lies in its low
nitrogen content and, in comparison to peat, the high proportion of easily rotting organic substances. In the decomposition of bark the micro-organisms combine with nitrogen and thus withdraw nitrogen from the surrounding soil. Therefore, in order to avoid possible deterioration of the soil and damage to plants,
about 1 to 2 kg of nitrogen per cubic metre of bark must be added. No proof has
been found up till now of the frequent allegation that there are substances in
bark which cause damage to plants; any such damage is generally due to this
withdrawal of nitrogen.

These difficulties can be avoided by composting the bark - the third form
of utilization - since during the rotting of the compost the proportion of
quickly decomposing material is greatly reduced and relatively large quantities
of the added nitrogen and phosphorus combine with the organic substance.

In the course of recent years various methods of composting bark have
been tested, since normally hardly any rotting takes place in large heaps of
bark. A certain degree of moisture, adequate aeration and suitable additives
are required before the micro-organisms can be activated. Preparations known
as compost starters, mostly consisting of certain cultures of bacteria, are now
commercially available. It has, however, been shown that with a proper dosage
of nitrogen bark compost can be ready in two to three months, or even in about
six weeks under favourable circumstances.

The main action of bark compost is not to manure the soil but to provide
it with a better supply of water and air, which both greatly promote root development.

Altogether it seems that there are favourable prospects for the sale of a
good bark compost, since there is an increasing demand for horticultural purposes. In addition, it must be borne in mind that, as a result of the activities of conservationists, the reserves of utilizable peat are becoming less and
less.

The difficulties attending the manufacture of bark compost should not,
however, be underestimated. A relatively large area of flat and easily accessible land on which it is possible for machines to move is required for stacking
the bark, while the cost of labour for turning over the stacks and loading and
packing the compost will be considerable. And for many firms the marketing of

the bark products could be their greatest problem.

For these and some other reasons, such as the provision of the necessary quantities of bark, it would appear advisable that the manufacture of bark compost and of bark particles for top soil treatment or for mixing with the soil should not be carried out in separate plants by individual firms but in a central installation operated jointly by several firms, as is done in Finland, for instance. For a single sawmill of medium size the additional investment and labour costs would probably be too great in relation to the quantity of bark produced.

Discussion

QUESTION: In my organization, we have sometime been working on the problem of utilization of waste, which are important problems to us. While your presentation has probably reviewed all possibilities of utilization, those of us needing to put it into practice find your suggestions difficult, particularly in respect of composting. In the United States wastes, particularly sawdust and bark, are used as humus and for soil improvement. I know there are institutes in the Federal Republic and elsewhere dealing with the problem, but these efforts often only involve publication of results. Development of actual techniques seems rare, is there anyone who can answer at all?

FRONIUS: While I think what you say has been true, the situation has recently improved a little in this respect. Practical work that I know of is going on in the German institutes and in at least three places investigation work is under way using bark as a waste product for humus, or composting. You referred to the problem of selling compost. I believe there are plans by Ministries in the different states of the Federal Republic to have a network of compost centers set up every fifty kilometers. Personally, I estimate the chances of other uses for the wastes as being greater.

VOLZ: I was in the western United States last year and I saw frequent use of bark to maintain moisture in the soil. It was also used to encourage plant growth along roadsides. In Scandinavia, bark has been burnt successfully for many years to produce heat for drying sawngoods. Could anyone from Sweden add anything on this?

HUNELL: Sawmills are becoming increasingly energy-dependent with development of production techniques, and at the same time, disposal of bark has become more problematical. Most of the world's energy production is about to undergo radical modification due to a whole series of problems connected with supply. Using bark to raise energy fits into this picture, though results do seem to be best when the bark is mixed with other combustibles.

ALOVIC, J., Faculty of Wood Sciences, Zvolen, CSSR: Though there are still many unknowns here, this does seem to be the direction for development.

HUNELL: Bark can be used to produce a variety of useful products, but none of these, in fact, are capable of consuming very large quantities of the substance. The main problem is the vast quantity of bark to be disposed of. Composting is practised to a limited extent in Sweden. It is also sometimes used for heating purposes or for drying, though with the oil crisis we had until recently, this was not very economical. Perhaps this particular situation will change. Bark

used in core layers of particleboards can help improve the board's insulation
quality, but this can equally well be improved by an air cavity. Whenever a
cheaper solution to that of using bark arises, bark is naturally rejected. One
possibility for the use of fairly large quantities of bark is in earthworking.
In Scandinavia we have problems with frozen ground, which can be prevented when
necessary by laying down a thick layer of bark.

FRONIUS: Thermal insulation properties of bark are, in fact, difficult to
utilize in particleboard. To obtain the rigidity usually required of these
boards, one has to increase density and soon one reaches the point where
insulation properties begin to deteriorate again.

Use of barking for protection of ground against heavy freezing has often
been tested in continental Europe but in these areas the long periods of severe
cold of the Scandinavian countries are seldom encountered, so this does not seem
to offer an opportunity for extensive use of bark.

May I now ask Professor Schulz, who has been involved in some of the work
we have heard described, to say something on the subject?

SCHULZ, H., Director, Institute for Wood Research and Technology, Munich, Fed.
Rep. Germany: The problem of bark utilization can be thought of as having
sprung up only very recently when it is compared to the thousands of years of
knowledge we have about wood itself. The current dilemma is to discover a few
of the basic properties of bark and to distinguish between the different types
of bark found in our part of the world. This was the first objective of the
studies carried out at the Wilhelm Klauditz Institute. Now, having reached this
stage of establishing the basic properties and distinguishing the bark types,
the utilization problem has so many aspects requiring separate research that one
institute is hardly able to deal with it all. Interest in carrying out this
work is developing, though, and a lot of work remains to be done.

KULTERER, H., J. Hasslacher Holzindustrie, Austria: At our mill we burn our bark
waste, but we have encountered a number of problems with it, giving rise to the
following questions: for how long can bark be stored before it loses its
combusting value, making it unsuitable for efficient burning? The second problem
is that bark with a moisture content greater than 50% is difficult to burn
effectively. Do you know of any practical experience on how this can be overcome, or is bark in this condition suitable only for disposal or composting?

FRONIUS: The question of storage I would like to pass on. I myself have found
that above certain moisture contents, bark begins to decompose and mould, and it
is certainly not possible to store moist bark for any extended time, it must
undergo initial drying, to some extent. Perhaps someone in the audience can
answer this more fully, or the other question concerning combustion of bark
with a humidity content above 50%.

THUNELL: Results of burning would depend considerably on the size of furnace
you have. I have been told by various suppliers that there are special furnaces
equipped with pre-drying systems that can solve the problem of very moist bark.
To keep bark in a combustible form over any period of time one has to follow the
same procedure as with chips and chip storage: it should be moved around
regularly with a suitable implement and not left in one pile, where moisture can
concentrate. It must be turned over from time to time. In this way you can

form at least a surge pile, certainly I see this frequently in Scandinavia.
Another alternative is to mix the bark with sawdust, or again, it can be
stored under roofing and kept heated.

13

Practical limitations to rationalization in sawmills: balancing investment costs with productivity

By Herr Dipl. Holzwirt Dr. Heinz Maisenbacher, Professor and Vice-President, College of Technology, Rosenheim, Fed. Rep. Germany.

The worldwide industrial expansion of recent years and recent decades has been characterized by ever-increasing mechanization and automation of the manufacturing processes.

The sawmilling industry has also felt the full effects of this process of industrialization. It has involved a rapid growth in capital requirements and of necessity a concentration of production capacity and capital - mainly outside capital in such proportions as to have, in some cases, jeopardized the profitability of the entreprise.

What is necessary therefore is to invest in a particular sector of the industry for which, from the provision of raw materials onwards, clear data are available. Investment should be allocated in such a way that the aims of all production processes i.e. the efficient manufacture of goods according to basic economic principles, be attained. To achieve these aims, every enterprise needs to undertake detailed analyses and planning with respect to the provision of raw materials, investment, production, finance and marketing. The greater the concern and the further the measures taken extend into the future, the more important and necessary this becomes.

Since decisions on investment always involve the long-term locking up of fairly large capital sums, a certain risk is entailed, against which the investment institutions will want to protect themselves. Apart from expert judgement, based on business experience, and intuition, modern management employs to an increasing extent systematic methods when it is a question of assessing the commercial efficiency of a plant and processes and the investment required.

Calculations of profitability consist mainly in analyses and comparisons of anticipated incomings and outgoings i.e. returns and expenditure.

Projected onto the conditions prevailing in the sawmilling industry, the following questions are extremely relevant to the required investment plans:

1. What investment plans are involved?
2. What are the possible alternatives?
3. What provision of capital is required?
4. What changes in costs and returns are to be reckoned with?
5. Upwards of what productive capacity or output do the measures
become economic i.e. as profitable as or more profitable than things
are at the moment?
6. After what period may one expect a reflux of capital?
7. What return from investment and from the resulting average yearly
surplus of receipts or the average yearly cost savings is to be expected, compared with the current situation or any possible alternative? If, before undertaking any investment, one ponders these
questions and tries to answer them as realistically as possible on
the basis of actual figures, then the steps essential to the responsible investor, if he is to protect himself against risk, have already been taken.

Investment planning

In the case of investment in plant and equipment, which is the only point
of discussion here, one can make a distinction in principle between

a) investment in the replacement of assets
b) investment in rationalization
c) investment in expansion.

Large-scale investment in replacement is comparatively rare in the sawmilling industry, and occurs in the case of relatively short-lived assets such as
vehicles or plant and machinery whose technical and economic life is exhausted.

In the case of investment purely in rationalization (when no increase in
production is possible, through lack of raw materials or through marketing difficulties), higher output for the same expenditure must be sought or the same
output with less expenditure. Frequently, however, investment in rationalization is associated with an increase in production, especially if it is applied
to bottlenecks in production.

Investment in expansion occurs primarily if a complete new mill is planned,
or else if investment in the replacement of assets inevitably involves investment in expansion, as in the case of framesaws, with their considerably higher
production capacity and the correspondingly higher investment.

With this allusion to the amalgamation of investment in replacement and
investment in expansion through technological advances, we are broaching the
entire problem which is the subject of discussion in all the countries with
any roundwood consuming industries.

Every instance of replacement investment that also involves expansion must
necessarily further the shrinking process within the sawmilling industry.

Investment alternatives

For large-scale investment, technology usually offers a number of possible
solutions.

Frequently the achievement of the highest technical efficiency is prevented
by the limited potentialities of the market, by shortage of personnel, and, finally, by financial bottlenecks.

For this reason it is important to work out alternatives and to tailor them
to the special needs and potentialities of the enterprise.

It is the primary task in calculating profitability to find the most appropriate alternative for the relevant enterprise.

Provision of capital

Each alternative solution demands a varying amount of capital. It is of
paramount importance for the economic success of the investment that this amount
should be taken fully into account.

As well as obligatory alternative offers by the relevant machine suppliers,
including their own anticipated production, alternatives should include considerations and calculations of additional financial investment arising from the increase in raw material and finished products stocks and from higher demand. Frequently one observes that it is precisely these elements that are neglected in
the case of expansion investment, or are not estimated with sufficiently realistically.

And, last but not least, in fixing the capital requirements for large-scale
investment in plant and equipment, attention must be paid to losses in production and to diminished output through the inexperience of personnel in the running in phase. Careful investors reckon, in the case of major changes in production technology, with starting-up periods of up to one year.

Changes in costs and returns

Understanding of the economic effects of investment enables one to compare,
as an aid to decision-making, the present and future ratios of costs to returns.
It is therefore necessary to take into account the costs which depend directly
upon investment, and also to allow for indirect changes in the area of costs
and returns.

In general one may proceed on the assumption that any investment results
in lower variable costs per unit of production. This advantage is counter-
balanced, however, by increased capital costs. In principle an investment plan
is only more profitable than the current situation if savings in the field of
variable costs cover the increase in expenditure. Whilst changes in costs connected directly with the project are still in the realm of the calculable, difficulties arise in estimating indirect changes in costs and returns, especially
if the investment is bound up with a considerable increase in capacity.

The additional capital costs already referred to are those resulting from
increased stocks of raw materials and finished products, and from the increase
of outstanding debts and perhaps from an extension of the repayment term due to
changes in client structure. But consideration must also be given to the effects upon the level of round timber costs and sales profits. Frequently cost
advantages in production must be sacrificed in part as a result of the higher
cost of raw materials.

The market situation and the opening up of new markets can often impose
a reduction in average sales profits. This is especially true when the intro-

duction of partly or wholly mechanised plant and equipment prevents a firm from
classifying its products by quality with the same care as before.

The reverse effects are also possible, however. This would be the case
when, as a result of higher capacity and new technology, goods are offered in
quality and quantity more in conformity with market requirements, leading to
smoother and more reliable service to the customer.

Moreover, in this connection it should be remembered that in the case of
highly mechanised installations with consequent increases in productivity, previous notions of mill utilization should be revised. Cases are known in practice
in which the total utilization of plant has been reduced by 4%.

Capacity and output

In calculating the economic effects of an investment and comparing with the
existing situation, costs should be broken into fixed and variable elements.

Take as an example the manager of a sawmill with a yearly output of
20,000 m^3 who is faced with the following choices: whether to

> a) continue to make use of the existing mill i.e. to make no invest-
> ment,
> b) make an investment for the modernization of the entire mill and
> equipment on the basis of a single framesaw, or
> c) to make an investment on the basis of two framesaws.

How should this manager proceed? Requirements are met by the following procedure:

> 1. Make an analysis of the costs of employing the existing mill breaking
> them down into fixed and variable elements. This gives the cost situa-
> tion of the existing mill.
> 2. Draw up planning data for the two alternatives, showing the machinery,
> installations and conveyor systems required, as well as any necessary
> constructional alterations.
> 3. Invite estimates.
> 4. Calculate investment in machinery and equipment and capital costs,
> bearing in mind the varying additional capital requirements for the fi-
> nancing of stock and outstanding debts.
> 5. Calculate the saving in time and/or personnel requirements in the
> alternatives chosen.
> 6. Compile a list of wage and social security costs, as well as variable
> costs directly connected with production, such as electricity consump-
> tion, lubricants, maintenance, repair costs, etc.

The following data are available from a practical example:
Existing sawmill (I)

> Variable costs: 35 DMk./m^3
> Fixed costs: 320,000 DMk./year

Modernized mill with 1 framesaw (II)

> Variable costs: 27 DMk./m^3
> Fixed costs: 450,000 DMk./year

Modernized mill with 2 framesaws (III)

        Variable costs: 22 DMk./m^3

        Fixed costs: 610,000 DMk./year

The following calculations can be made from this simplified example:
Per-cubic meter production costs of the existing sawmill I are equalled by sawmill II when annual production reaches 16,200 m^3, and by sawmill III at
22,300 m^3. The same marginal capacities can be derived mathematically using the
following formula, using x for marginal capacity, and C for cost (f for fixed
and v for variable)

$$C(x) = C_{fI} + xC_{vI} = C_{fII} + xC_{vII}$$

This can be rearranged to give a figure for the marginal capacity in sawmill II:

$$x_{III} = \frac{C_{fII} - C_{fI}}{C_{vI} - C_{vII}} = \frac{450,000 - 320,000}{35 - 27}$$

$$= 16,000 \text{ m}^3 \text{ (approx)}$$

Similarly the marginal capacity of sawmill III (as compared with sawmill I) can
be obtained:

$$x_{III} = \frac{610,000 - 320,000}{35 - 22} = 22,300 \text{ m}^3 \text{ (approx)}$$

Note that this comparison does not take into account any rise in raw
material costs nor any reduction in machinery availability.

If one were to estimate for these extra costs a total of 4 DMk./m^3, the result
for sawmill III would be a marginal capacity (compared with I) of about 32,200 m^3.
If one compares sawmill II and sawmill III, then the marginal capacity of III
works out as

$$x_{III} = \frac{610,000 - 450,000}{27 - 22}$$

$$= 32,000 \text{ m}^3$$

Return on capital - amortization

    Potential production capacities and the levels of production required for
profitable operation can be estimated when technical design details are established and costs calculated. Allowances must be made for market situations for
raw materials and finished goods. For an investment to be viable this calcula-

tion must yield either a surplus in income or a saving in costs.

The theoretical amortization period can be calculated from the relation of invested capital to mean annual cost savings or mean annual surplus. This is the period after which the invested capital is repaid and is no longer at risk. It corresponds to the shortest useful life of the investment.

In the procedure we describe, it is up to the investor to decide what period of amortization he regards as the criterion for acceptable investment.

In other branches of industry it is known that large companies work to priorities on this and give preference to investments with amortization index figure of 3 or less. In the sawmill industry the figure sought is well above this, except in cases of small-scale rationalization.

Two methods are normally used to present and compare alternate choices as an aid in decision-making before an investment is made. They can be termed the static and dynamic processes.

Static processes use average costs, that is, costs and return figures that are constant year by year. This process takes no account of variations in the incidence of costs and returns during the period involved.

Dynamic processes take account of the influence of time through interest on payments. In the simplest of cases, calculation of the period of amortization for example III was carried out as follows:

Capital invested ....1.8 million DMk.
Useful life of installation ....10 years
Actual conversion capacity ....25,000 m^3
Annual saving in costs
compared with the existing mill...........................115,000 DMk.
Offsetting of losses on operating hours
against additional profit per year........................–50,000 DMk.
Depreciation per year.....................................180,000 DMk.
Surplus receipts per year (before taxes)..................245,000 DMk.
Theoretical amortization period ....7.35 years
Remainder of useful life ....2.65 years
Surplus receipts during remainder of
useful life (before taxes) ....650,000 DMk. (approx)

In a survey covering several years on the assumption of a 10% annual increase in costs in the field of variable costs, the calculation of profitability for plant III, summarized by comparison with plant I, would appear as it is shown in the accompanying table.

Returns

Calculation of returns involves determining the earnings capacity of the capital invested from the average yearly surplus of receipts or the average yearly savings in costs, in relation to the capital invested.

When examining the profitability of an investment, the objective is to compare investments in machinery and equipment with financial investment, and

Comparative calculation for plants I and III

Figures in 1,000's of DMk.

Year	1	2	3	4	5	6	7	8	9	10
Annual cost savings compared against I	45	91.25	141	194.5	252	313.75	380.25	452	554.25	613
Difference between operating hour losses and additional profits through increased production	50	50	50	50	50	50	50	50	50	50
Profits before tax	5	41.25	91	144	202	263.75	330.25	402	504.25	563
Profits after tax (approx.)	5	29	45	72	101	132	165	201	252	331
Depreciation	180	180	180	180	180	180	180	180	180	180
Surplus receipts	175	209	225	252	281	312	345	381	432	511
Loan at start of year	1,800	1,625	1,416	1,191	939	658	346	1	–	–
at end of year	1,625	1,416	1,191	939	658	346	1			
Reserves	–	–	–	–	–	–	–	380	812	1.323

to invest one's capital where it yields the highest return. Here, however, one must take into account the drop in value in investment in plant and equipment. It is therefore necessary to relate the annual surplus of receipts of an proposed investment to the average book value of the plant or to the book value at any given time.

In tnis connection it must be pointed out that, especially in the calculation of returns, the use of compound interest makes more accurate allowance for the periodical accumulation of yields from investment than does the static method, which works on the basis of average cost savings and average surplus receipts.

Conclusion

Investment in the sawmilling industry now involves far greater capital commitments than normal in the past and will continue to do so. It is therefore essential, before making major decisions regarding investment, to employ an appropriate method, which, free from amateurishness and prestige considerations, can provide an answer to the following questions: How profitable is the investment? Within what period can one count upon a return of capital? What return can be expected?

Although the final decision on the placing of an investment cannot be made solely on the basis of a calculation which rests partly on figures lying in the future, the calculation of profitability represents an important aid. It has on the one hand frequently protected the investor from surprises and unnecessary losses, but on the other hand has certainly just as often convinced him that an investment was desirable, necessary and accompanied by less risk than at first appeared.

By laying bare risks of investment and setting down operational parameters for the proposed mill the profitability calculation proceedure is an aid to decision making for the investor. Bringing about the type of operation described is the first task of all those involved in the enterprise after the investment is made.

Discussion

BRAUN, France: I am a little sceptical regarding the amortization period you quote and, in particular, regarding the working life of machines. With the high machine operation speeds obtained in modern mills, I do not think that the operating life of ten years that you refer to is realistic, today's high-speed machines do not seem to last that long, and therefore the amortization calculation has to be modified.

MAISENBACHER: This is a significant point since depreciation costs could rise with output and render unviable a project that would seem economic, were it assumed to have a longer machine-life.

I wanted to emphasize how essential it is that investment calculation must be made in a consistent way if decisions are to be based upon them. In modernizing a sawmill, data must be available to allow an understanding of the entire operation, and this data must be kept available to allow the mill, once built, to be managed properly. The specific data must relate to costs, turnover and markets.

NJOTO, T., Kayu Mas, Jakarta, Indonesia: Can you give me a figure for the level of investment required for a sawmill to produce 100 m^3/hr with German machines with Meranti timber as the main product?

MAISENBACHER: With this type of timber, assuming a recovery factor between 50 and 55%, you would require to use 200 m^3/day runwood intake, i.e. 40.000 m^3/year. For a sawmill to convert this, you would have to calculate in the order of magnitude of DMk 7 to 8 millions for machines, buildings, the entire plant.

NJOTO: I assume this figure takes into account normal European conditions, i.e. a large building, electric supply, and special systems, many of which would not apply to our case in Indonesia. The mill we are speaking of would not use computers, and would only be semi-mechanized, certainly not fully automated.

MAISENBACHER: This is quite difficult to answer. One would have to know more about machine requirements and just how rudimentary a plant you have in mind. Even with this machinery, one would have to know something about the degree of mechanization you require. Do you want conveyors, and what type of conveyors? If the requirements are only for a rudimentary design, it would be possible to think of a sum of around DMk 2 million.

QUESTION: In view of what Mr. Braun said about raw material shortages, it would be interesting to hear your reactions, Herr Maisenbacher, to the idea of a mill intentionally reducing production in a combined effort to overcome the problem, and to know how you think its operation would compare with that of a mill undertaking expansion.

MAISENBACHER: Modernization is a process going on continually in the sawmilling industry, and this has to remain so unless you are going to pass laws against technological progress. The question of increasing production arises with almost any mill undergoing modernization in view of the large amount of capital involved. You are, of course, right. It would be nice if sawmills could subsist just on improved technology without any increase in production capacity. But unless calculations, accurate calculations, are made in advance, many sawmill operators discover early after the investment that the only way to improve a cost-covering operation is by going for higher production, which can then give a modest profit.

14

Forecasts for the sawmill industry in Europe: development trends and effects of modernization

By Mr. Henri J.R. Widmer, FAO (Food and Agriculture Organisation of the United Nations) Forest Industries and Trade Division, Geneva, Switzerland.

The task of looking into the future to forsee what lies ahead is not made any easier by the present unusual circumstances, which have been said by many to indicate that an era is coming to an end. It is not only the sudden awareness that man's existence is ultimately dependent on raw materials (even though the form of his existence is determined by the use he makes of them) nor that the supply of raw materials, at least many of these that are exploited intensively and at progressively increasing rates, is limitless. Other, economic as well as social forces are also changing the face of society and thereby of the economy.

Although wood is a renewable resource, and may well be favoured by future developments, predictions over the longer term appear particularly hypothetical at the moment. It is certain, however, that not only at the global level, but also in the narrower field of forestry and wood, things are changing, at the technical and economical levels; one visible sign is the new price relations which have brought about a much needed correction both for the wood processor and the forest owner. For these reasons, the emphasis in my paper will be on reviewing trends and those underlying factors that I consider relevant for the sawmilling industry, rather than on predicting the future.

Mecanization in the sawmilling sector

In recent years, modernization in the sawmilling sector has progressed substantially, and this year will see a further acceleration. One could say that sawmilling has entered the industrial age, after having for a long time retained many of the characteristics of an artisanal activity, although the forest-rich countries have had some large sawmills for a long time. For many years it was considered a technically backward industry with a marginal growth of output and

a record of modest profitability, but now the emergence of the industrial sawmill
is rapidly changing this image and making the industry as a whole - if you will
pardon the expression - respectable to the investor.

The effects for the sawmill of these developments are essentially of three
sorts: (i) rationalization, with an increase in productivity, but necessarily an
increase in capacity; (ii) specialization; and (iii) increase of capacity. In addition they have broadened the raw material base of the industry in the direction
of small-sized roundwood and offer possibilities to improve yield and the quality of the final product.

The rise in productivity is quite spectacular. In Austria, it has reportedly been of the order of 80% between 1961 and 1972. Average annual output per employee rose from about 270 m^3 to nearly 500 m^3. In a recently built sawmill in
Finland, which is fully automated, the capacity is about 150,000 m^3 and it is
said to be the most modern in Europe, output per employee is around 4000 m^3.

Any decision on investment in new machinery equipment for modernization,
specialization and possible further processing naturally requires careful consideration, as it is an operation that implicitly affects every stage of the
production process. For instance, increased speed at the saw has to be balanced
by corresponding adaptations at the preceding and following stages. Where mechanization results in substantially increased capacity - and this may be the case
more often than not, as modern equipment is designed to give high speed and output - particular attention must be given to questions of raw material availability and of markets.

Another point that assumes added importance is waste utilization, particularly as some of the modern processing techniques tend to produce a higher percentage of sawmilling waste in the form of chips. This waste is, however, usable
in other industries, without further intermediate processing. The economics of
this may be most marked in the case of directly integrated mills.

The structure of the sawmilling sector

An idea of the structure of the sawmilling industry is given in Table 1
(see page 138). It gives relevant data for a few countries which can be assumed
to be fairly representative of much of Europe, although in Eastern Europe concentration into larger units has pursued vigourously, in line with general industrial policies.

Two facts are immediately obvious: firstly the large number of small sawmills, and secondly, their small share in total production. As a general rule,
the small sawmills serve more the local demand and sometimes work intermittent-
ly or on a seasonal basis. It is, on the other hand, interesting to note that a
varying, sometimes appreciable, number of sawmills are associated or integrated
with other activities. In the case of very small sawmills there are sometimes
agricultural or other rural activities. In France, integration with logging operations is not uncommon. For other, medium or larger sawmills, integration is
common with woodworking industries which use sawnwood (e.g. window manufacture,
joinery) but these mills can also be linked to timber trading activities. Integration with panel, pulp and paper industries is more common for large mills. In
Finland, over half of the sawnwood was produced in 1972 by enterprises which had
pulp, paper and wood-based panels interests as well.

What have the trends been up to the early 1970's? The number of sawmills tended to decrease, in some countries by up to a third since the 1950's, while the average size increased. Overall capacity seems to have tended to increase, not only in countries where forest resources easily permitted further expansion, but also where domestic wood availability would have appeared to be a limiting factor, and where capacity was sometimes already under-used. Integration has been increasing with sawn wood-using sectors, and notably in the Nordic countries with industries using small-sized wood and sawmill residues. This is having a substantial effect on the economics of raw material utilisation. The use of residues will be of crucial importance both for the individual sawmill as for the wood sector as a whole.

Table 1 - European sawmill industry
          structure, selected countries

(approximate figures relating to 1970 unless otherwise stated)

| Countries | Total number of sawmills | of which with a production | | | | | |
| | | up to 1,000 m³ | | 1,000 - 5,000 m³ | | over 5,000 m³ | |
		number	share of production %	number	share of production %	number	share of production %
Austria	4,000	3,500			. . 35%*	500	65%*
France	7,000	4,800	16%	1,800[1]	37%	400[2]	47%
Finland	8,770[3]	8,600			. . 16%	170[4]	84%
Sweden	3,500	2,600[5]	5%	900[6]		. . .	.95%
* Estimate							

1/ 1,000 - 4,000 m³   2/ over 4,000 m³   3/ year 1972   4/ of which the ten largest (with over 125,000 m³), accounted for about 24% of total output
5/ up to ,400 m³   6/ over 1,400 m³

Trends on the forest and on the market side

On the forest side, apart from the general question of wood raw material
availability to which I shall come later, the sawmilling industry has to take
account of the rationalization process in forest management and in logging. Different forms of cooperation between private forest owners have been developed,
and, if they have not generally gone as far as in the Nordic countries, where
forest owners' organisations have set up their own in forest industries, the
question is nevertheless posed as to whether they should take a greater interest
and have a direct influence on the marketing and utilization of their timber.

In the logging sector, one important development is the central conversion
sites which are not only a means of mechanizing harvesting operations but also
to some extent a tool for roundwood marketing.

On the market side, we also discern a number of new developments, which,
if they have not yet attained the same degree of intensity, yet, will nevertheless inevitably affect the entire sector.

Firstly the structure of the trade itself and of the traditional pattern
of distribution, where the different functions in the chain were more or less
clearly defined, is changing in our particular sector as they have done in others.
The trend to larger enterprises with horizontal and vertical ties across the sector can improve to a considerable extent the economics of sorting, handling,
stockkeeping etc. of timber and shorten the trading circuit.

At the same time the range of available wood products, e.g. tropical species and panels, has become much larger, although this also applies of course to
materials capable of substituting wood. This fact, and the awareness that they
need to defend their own position and that of their products effectively has induced producers in a number of exporting countries to establish closer contacts
with users and consumers in importing countries.

Nowadays, sawnwood itself is no longer the commodity it used to be, but
has become more of a technical product. If it has in the past been able to maintain its position better than expected in many end-uses where it was exposed to
strong competition, this is due to its inherent qualities, but probably as much
to intensive research and development work through which it has been possible to
improve its performance. Modern gluing and fingerjointing techniques which permit the production of almost any desired dimension, shape and quality of sawnwood from almost any type of roundwood, are two examples.

In specification and grading, more and more attention will be paid to the
intended end-use of the pieces, an example being stress-grading, which has been
dealt with already in detail.

Considerable efforts have been made in the field of standardization, although much remains to be done at the practical level. Progress is slow despite
the fact that studies, carried out in Sweden, show standardization in the number
of assortments produced - these can be several thousand in a medium-sized sawmill
between species, qualities and dimensions - are preconditions to obtain the full
potential benefits of sawmill rationalization.

One reason for this is undoubtedly the diversity of end-uses for sawnwood,
and the small degree of standardization achieved in some of these end-uses, e.g.
building. It is contended that in many countries house owners insist on individual design in spite of the high cost level which building has attained. It is
also said that the very fact of producing all the different assortments constitutes a competitive advantage over producers e.g. in North America. It would
seem, however, that the advantages to be gained from a more standardized production, on the forest side as in processing and utilization, are so considerable
that efforts should be intensified and concerted in this direction.

Balance of roundwood supply and demand

After these more technical aspects, let us now turn our attention to the
demand and supply for sawnwood and its raw materials in Europe. Table 2 (see
page 142)summarises the developments between 1950 and 1970 and shows the overall wood balance in the sectors that interest us. It further contains tentative
forecasts for 1980 and 1990. These forecasts, as well as the other data are based
on studies carried out on timber trends and prospects by the UN/ECE Timber Committee, the most recent of which was published in 1969 under the title: European
Timber Trends and Prospects, 1950 - 1980, An Interim Review. The forecasts set
out in Table 2 of this paper have been modified somewhat from those in the Interim Review, to take into account subsequent developments. I may add that our forecasts have generally proved to be fairly accurate, though somewhat too low notably in the case of sawn softwood and particleboard. I would advise you, however,
not to keep your eyes fixed on the absolute value of the figures in the forecasts,
but rather on the trends they indicate. I should add that one of the underlying
assumptions was that no major disruption in economic activities would occur and
furthermore that price relations would not change radically. Also, the forecasts
were made before the oil crisis.

What then do the data show us? Firstly, the importance of sawnwood in the
forest products sector, and, implicitly, that of sawlogs in forestry. When all
European removals are taken together - including firewood and such industrial
roundwood as pitprops and poles - the share of sawlogs in the total was 34% in
1950 (out of 293.8 million m^3), and 42% in 1970 (344.7 million m^3).

Production of sawn softwood increased roughly in line with removals of coniferous sawlogs, but that of sawn hardwood rather faster than removals of hardwood sawlogs, due mainly to the availability of substantial and continuously increasing volumes of tropical hardwood logs.

Trade in sawnwood followed a different pattern, as net imports in sawn
softwood grew rather faster than in the case of sawn hardwood, although in recent
years imports of sawn hardwood from tropical countries have tended to accelerate
their growth.

Consumption therefore of both sawn softwood and sawn hardwood increased
substantially with net imports attaining 8.11 million m^3 for sawn softwood (re-
presenting 13.54 million m^3 in terms of WRME = Wood Raw Material Equivalent), and
1.23 million m^3 for sawn hardwood (2.24 million m^3 WRME), or 11.3% and 6.4% respectively of apparent consumption.

Table 2 - European sawlog and pulpwood
          balance and forecast

Million WRME[1]				
	Actual			Forecast	
	1950	1960	1970	1980	1990
Products of sawlogs					
Apparent consumption	103	138	172	198	210
Removals	100	118	145	165	180
Net imports	+3 [4]	-20	-27	-33	-30
Products of pulpwood					
Apparent consumption	37	74	142	245	390
Removals	37	62	95	160	220
Residue transfer	5*	13	27	40	60
Net imports	+ 5 [4]	+ 1 [4]	-20	-45	-110
* Estimate					

European sawlog and pulpwood products balance, 1950 to 1970, and tentative forecasts for 1980 and 1990, excluding firewood and certain roundwood assortments mainly used in the round.

   1/ Wood Raw Material Equivalent, i.e. the volumes of processed products are expressed in terms of the corresponding volumes of raw wood (roundwood, and residues) required to produce them.

   2/ Tentative forecasts revised against those made in the last FAO/ECE Study on European Timber Trends and Prospects 1950-1980, an Interim Review (published 1960)

   3/ Higher of two alternative levels

   4/ Net exports

The overall growth in apparent consumption can roughly be divided into that
due to the growth in population and an increase in per capita utilisation. Thus
for sawn softwood, about 2/5 of the total increase was due to the growth in population, and 3/5 to an increase in per capita consumption (from 117 to 145 m^3 per
1000 inhabitants between 1950 and 1970); in the case of sawn hardwood, close to
1/4 of the total reflected growth in population, but just over 3/4 was due to an
increase from 25 m^3 to 39 m^3 per 1000 inhabitants in apparent consumption.

What are now the prospects for the demand and supply of sawnwood and sawlogs, and their balance in the medium and longer term? I have already said that
the relevant figures in Table 2 should be considered as tentative estimates. Detailed forecasts up to the year 2000 will be elaborated in the new Timber Committee study on timber trends and prospects work on which has started earlier this
year and which will be terminated later in 1975 or early 1976. Special emphasis
will be given in this study to the questions of raw material availability and
forestry.

As far as future consumption is concerned (not to be confused with demand)
the Interim Review assumed that per capita consumption of sawn softwood would not
increase further, till 1980, in spite of an expected further growth in Gross National Product and per capita incomes, while for sawn hardwood a further, though
modest, increase in consumption would come in the first instance from population
growth. Taking all products derived from sawlogs together, i.e. including plywood and veneers, which however account for a rather small share of the total,
Table 2 shows us that we may expect a further, though more moderate growth in
consumption, and an increase in removals which will not, however, equal growth
in consumption, so that net import requirements should show an increase in 1980
over 1970 of about 6 million m^3 WRME, or over a fifth.

In the sector of pulpwood and its products, consumption is expected to increase much more rapidly, while removals, and residue transfer will also increase but more slowly in terms of volume, so that net import requirements would more
than double between 1970 and 1980 and be substantially higher than those of sawlogs, veneer logs and their products.

A few words of comment on removal estimates in Europe are called for. These
estimates are based on the latest data available as provided to the 16th session
of the FAO European Forestry Commission in 1972. They show that in a number of
countries, e.g. Austria, the Federal Republic of Germany and Switzerland, recent
inventories had indicated that growing stock and increment were appreciably higher than 10 or 20 years ago. There is also no doubt that forest production, notably in small private forests, could be further increased by more intensive silviculture, fertilisation, quick-growing species, better varieties, etc. A further
point to note is that agricultural land has been abandoned at an increasing rate
in Europe over the last two decades, and this could be available for forestry
purposes. A further possibility to increase raw material availability would be
by making fuller use of the whole tree - what is currently used represents generally less than 2/3 of the tree volume (including bark and needles).

However removals, especially by small private forest owners, are influenced
not only by market conditions, but by many other, not necessarily purely economic
considerations such as fiscal law, the size of forest holdings etc. I would like
to draw your attention here to the very small size of private forest holdings in

many countries in Western Europe, as compared to Scandinavia where they average some 30 to 40 ha.

Increases in forest production apart, let us not forget that other factors have a limiting effect on wood production, and here I am referring especially to the demands for indirect and social benefits from the forest which will undoubtedly assume even greater importance in the future.

We have noted that the gap between Europe's own roundwood production and its consumption of wood products will widen further, in other words, its dependance on imports from other regions will increase. The question is therefore: will these quantities of wood be forthcoming, and at what price? Although a simple answer to this question cannot be given, the existence of huge forest resources in the tropics, in North America and the USSR, many of which are still virtually untouched means that much larger quantities of wood than Europe could require over the next two decades, can be made available. No doubt the utilization of these forests poses many problems, e.g. through the distances involved, the particular topographical and climatic conditions, the heterogeneity of the tropical forest, but technically these can be solved, and the price levels attained recently - which in the ultimate analysis reflects a more realistic valuation of the material - could well act as a stimulus for exploitation in these regions. It is probable that the greater part of this would be exported in a processed form.

An attitude of complacency as regards their future, therefore, seems inappropriate for those in the wood industries and forestry in Europe. On the contrary, continued efforts will be necessary to rationalize at all stages, to improve raw material utilization and the performance of the product.

It is quite possible, of course, that the net wood deficit in Europe could lead to increased competition between different wood industries, notably the sawmilling industry and the pulpwood (or small-sized industrial wood) using industries, the more so as the new techniques enable the sawmills to accept small diameters not used previously.

Conclusion

Certain deductions can be made from the above:

Demand for the products of the sawmilling industry can be expected to grow more in Europe, though the rate of growth could be somewhat lower than in the past decades. This forecast is based on the assumption that no major disruptions occur on the economic scene. Availability of sawlogs - notably of coniferous species - from European forests will also increase, in some countries somewhat faster than before. In addition, the new sawmilling techniques can broaden the raw material base of the industry in the direction of smaller dimensions. The availability of these latter is also expected to increase partly on account of thinnings which have been delayed in recent years for economic reasons.

European forests will, however, be providing a diminishing share of the raw material for the region's expected consumption of sawnwood and even more of certain other products made from wood. While it can safely be assumed that the resulting deficit will be covered by supplies from other regions, probably in processed form, competition for roundwood between different industries could increase.

The sawmilling industry will also be faced with further developments both on the forest and market sides, e.g. further moves towards concentration in one form or another, as a consequence of those economies of scale resulting from technological progress.

If the main underlying factors appear fairly clear, at least in a qualitative sense, it is less easy to assess their effect on an industry such as sawmilling where the owner is generally also the manager, and is responsable for the enterprise. His flexibility and ingenuity are one of his most important assets in solving the problems that he will face. And the problems will be considerable.

While it is obviously necessary to rationalize continuously, the raw material situation must also be considered carefully, as the level of investment necessary to this end is considerable and presupposes adequate capacity utilization. The same can be said for the market. I imagine that some of the problems of the future could be best solved by establishing some form of cooperation between individual sawmills, e.g. for their production and marketing policies.

Kapitel 5/Chapitre 5/Chapter 5 — Zeiten/Pages/Pages 38 — 55

1. Seite 56 - Fig. 1 — Der Planschnitzler hat sich in der europäischen Sägewerksindustrie mit mehr als 150 Maschinen sowohl als Schwachholzmaschine als auch als Reduziermaschine behauptet.

1. Page 57 - Fig. 1 — La découpeuse à surfaces planes a fait ses preuves en Europe pour les petites grumes aussi bien que pour la réduction des plans.

1. See page 50 — Fig. 1 — The canter has achieved a leading position in the European sawmill industry both for small logs and as a face cutting reducer. Some 150 are in use.

Kapitel 6/Chapitre 6/Chapter 6 — Zeiten/Pages/Pages 56 — 65

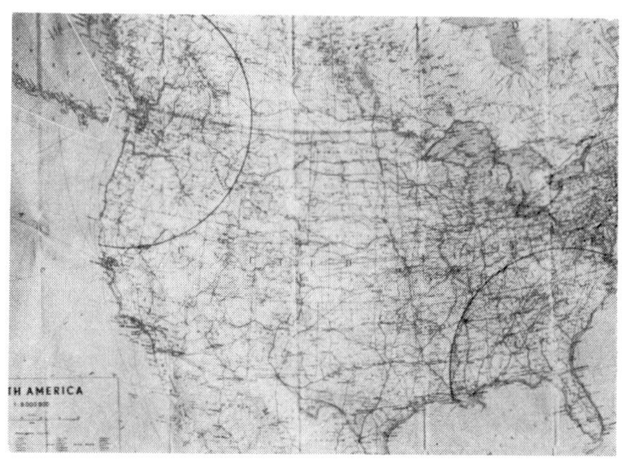

2. Seite 61 — Auf dieser Reise werden wir die westlichsten Teile der Vereinigten Staaten und Kanadas einblenden. Der Süden ist ein weiteres Gebiet wachsender Bedeutung.

2. Page 64 — La partie dont il est question est la région très boisée qui se trouve à l'extrême-ouest des Etats-Unis et du Canada.

2. See page 56 — The far western section of the USA and Canada is the main timber-growing area to be dealt with in the talk. The South is an area of growing importance.

3. Seite 62 — Ein gummibereifter Skidder. arbeitet in einem Bestand von Murraya-Kiefern, typisch für den Sekundärwald, der im Nordwestpazifik-Gebiet, im Südosten der USA und im Osten Kanadas zu finden ist.

3. Page 64 — Engin à pneus en usage dans un lot de pins "Lodgepole" aussi utilisé au sud des E.U. et dans l'est du Canada.

3. See page 57 — A rubber-tyred skidder works a stand of lodgepole pine typical of second-growth pine found in the Northwest, the southeastern USA and Eastern Canada.

4. Seite 63 — Die Grösse dieses Sägewerkes mit entsprechendem Sammelplatz im südlichen Teils von British Columbia ist typisch. Das Sägewerk befindet sich rechts hinten.

4. Page 65 — Scierie de grande taille en Colombie Britannique. On remarque l'ampleur de l'aire d'empilage. La scierie se trouve à l'arrière-plan à droite.

4. See page 58 — Large size of this mill in the southern interior of British Columbia is quite typical, with a log deck sized to match. Saw house is at the rear, right.

5. Seite 64 — Die grosse Kapazität des Staplers ist hier klar ersichtlich. Er sammelt Stämme, die in Form von Flössen vom Oberlauf des Columbia-Flusses heruntergebracht werden.

5. Page 66 — L'ampleur des aires d'empilage est évidente. L'engin ramasse des grûmes descendues en radeaux sur la rivière Columbia.

5. See page 58 — Large size of the log stackers used is evident here. This machine is gathering logs brought down the Columbia river in log rafts.

6. Seite 64 — Innenansicht eines grossen amerikanischen Sägewerkes zeigt, wie alle in Arbeit befindlichen Gegenstände auf Rollenlager oder Förderkette transportiert werden und nie in Hubwagen.

6. Page 66 — Vue intérieure d'une usine type des E.U. montrant l'usage de chaînes transporteuses plutôt que de chariots élévateurs ou à main.

6. See page 59 — Interior view of typical, large US mill shows how most work-in-progress is transported by rollcase or transfer chain and never by handcarts or lift trucks.

7. Seite 64 — Für grosse und mittlere Stämme benutzen die meisten Werke im Westen Nordamerikas Bandsägen wie diese 3-Meter Doppelschnitt-Bandsäge. Der Mann in der Mitte transportiert das Material ab.

7. Page 67 — Pour les grûmes grosses ou moyennes on utilise souvent une scie à ruban de 3 m. à double lame. L'homme au centre est le déchargeur.

7. See page 59 — For large and middle-sized logs most Northwestern mills use a bandmill headsaw such as this 3-meter double cut saw. Man in centre is the offbearer.

8. Seite 64 — Beliebt ist in Nordamerika für den ersten Vorschnitt kleiner Stämme der "Beaver" oder "Chip-N-Saw", der die äusseren Schwartenflächen nur abspant. Grössere Maschinen machen auch innere Schnittlinien.

8 Page 67 — Scieuse-fragmenteuse pour le débitage préliminaire des petites grûmes évitant la perte de sciure par trait de scie extérieur.

8. See page 59 — The chipping headrig is a popular approach to initial breakdown for small logs. This saves the outermost sawkerf loss. Larger units also make interior saw lines.

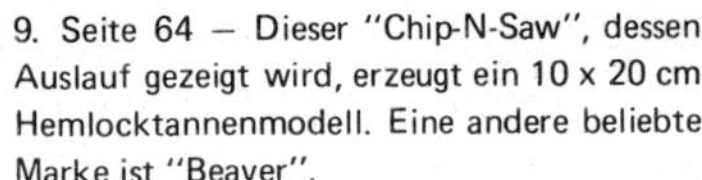

9. Seite 64 — Dieser "Chip-N-Saw", dessen Auslauf gezeigt wird, erzeugt ein 10 x 20 cm Hemlocktannenmodell. Eine andere beliebte Marke ist "Beaver".

9. Page 67 — Sortie de "Chip-n-saw" montrant une pièce de sapin-cigüe de 10 x 20 cm. Une autre marque souvent utilisée est le "Beaver".

9. See page 59 — This Chip-n-Saw, of which the outfeed is shown, is producing a 10 x 220-cm hemlock cant. Beaver is another popular make.

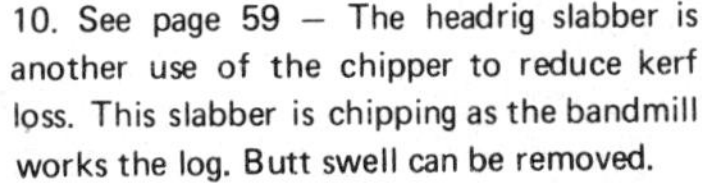

10. Seite 65 — Die abspanende Brettholzsäge erlaubt Fugenverlust zu beschränken. Diese Säge spant ab, während die Bandsäge gleichzeitig den Stamm bearbeitet, Die verdickte Basis wird leicht entfernt.

10. Page 67 — Autre moyen de réduire la perte en sciure est une fragmenteuse qui fait les copeaux pendant que la scie à ruban découpe la grûme.

10. See page 59 — The headrig slabber is another use of the chipper to reduce kerf loss. This slabber is chipping as the bandmill works the log. Butt swell can be removed.

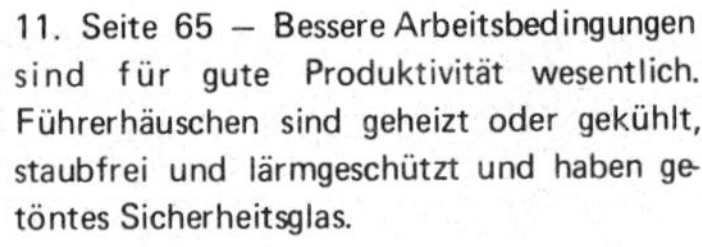

11. Seite 65 — Bessere Arbeitsbedingungen sind für gute Produktivität wesentlich. Führerhäuschen sind geheizt oder gekühlt, staubfrei und lärmgeschützt und haben getöntes Sicherheitsglas.

11. Page 68 — L'amélioration des conditions de travail permet un meilleur rendement. Les cabines climatisées ne laissent pénétrer ni bruit ni poussière.

11. See page 60 — Improvement of working environment is considered essential for good productivity. Operator cabs are temperature controlled, noise- and dustproof.

12. Seite 65 — Die meisten amerikanischen Besäummaschinen laufen auf Doppelachsen, wie diese Schurman-Maschine, deren Auslauf wir hier sehen.

12. Page 68 — La plupart des chanfreineuses U.S. sont à arbre-double, tel cet engin, type "Schurman" vu à la sortie.

12. See page 60 — Most American edgers are of the double-arbor type, such as this Schurman machine seen from the outfeed.

13. Seite 65 — Dieser Gatterbesäumer ist 152 cm breit, Sägen mit 55 cm Durchmesser werden benutzt. Der Auslauf ist auf diesem Bild zu sehen.

13. Page 68 — Chanfreineuse de 152 cm. avec scies de 55 cm. de diamètre et trait de scie de 3,1 mm, vue de la sortie.

13. See page 60 — This edger is a 152-cm wide unit using 55-cm dia saws with 3.1-mm kerfs, seen from the outfeed side.

14. Seite 65 — Die Doppelbandsäge ist eine der im Westen benutzten Typen und wird stets beliebter. Sie kann jedes Stück mit einer flachen Seite bearbeiten und erzeugt weniger Fuge.

14. Page 68 — Dédoubleuse souvent adoptée dans l'ouest. Peut traiter toute pièce ayant déjà un côté scié, avec une lame plus fine.

14. See page 60 — The twin band resaw is one of several types being adopted in the West, and is rapidly gaining favour. It can cut any piece with a flat side with less kerf.

15. Seite 65 — Ablängsägen sind gewöhnlich Vielblattkreissägen, in die Stücke einzeln eingeführt werden, mit einem Speichersystem, das die richtige Säge fallen lässt, sobald das Stück ankommt.

15. Page 68 — Les équarisseuses sont généralement à scies circulaires multiples où les pièces passent une par une et la scie est actionnée électroniquement.

15. See page 60 — Trim saws are generally multiple circular saw units into which pieces are fed singly and with a memory that drops the required saw when the piece arrives.

16. Seite 65 — Schnittholzsortierer mit automatischen Speichersystemen sparen viel manuelle Arbeit. Die Sortierungen erfolgen nach Breite, Länge und Qualität.

16. Page 68 — Plateaux de triage à bascule avec mémoire. Ce système résulte en économie de main-d'oeuvre. Le triage porte sur largeur, longueur et qualité.

16. See page 60 — Tray sorters with unloading tipples and automatic memory system have saved much manual labour. Sorts can be made to width, length, and grade.

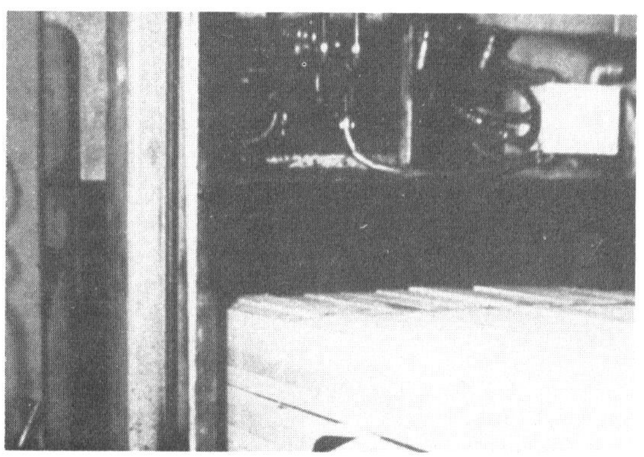

17. Seite 66 — Ein grosser Teil der Stapelung ist automatisch, wie auf diesem Verpackungsstapler der Firma Moore.

17. Page 68 — L'empilage est automatisé. Ici une empileuse-conditionneuse automatique "Moore".

17. See page 60 — Most strapping is automatic, carried out as with this Moore automatic strapper.

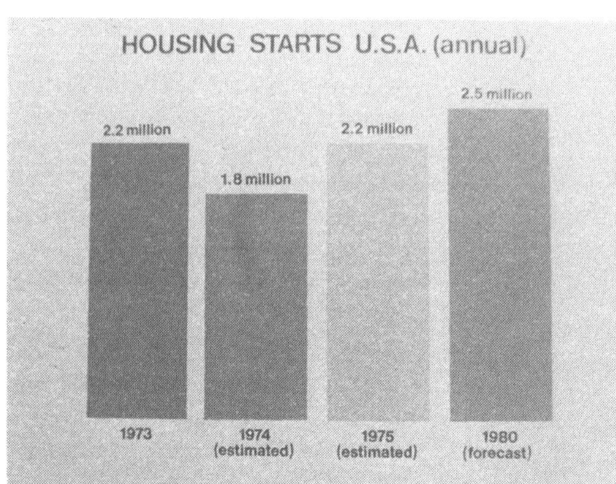

18. Seite 66 — Die Zahl von Wohnhäusern beeinflusst in den USA die Holzindustrie wesentlich, da soviel Häuser aus Holz gebaut werden. Diese Tabelle zeigt jährliche Statistiken.

18. Page 69 — La construction d'habitations, principalement en bois aux E.U. est d'une grande influence sur le marché. Tableau de statistiques annuelles.

18. See page 60 — The number of housing starts in the US is a major influence on the lumber industry, since so much housing is built with wood.

19. Seite 66 — Dieser Morbark Spanernter bearbeitet den ganzen Baum : Hauptstamm, Äste und Blattwerk, und hilft daher, den wachsenden Bedarf an Fasern für die Zellstoffindustrie zu decken.

19. Page 69 — L'abatteuse-fragmenteuse "Morbark" traite l'arbre entier, tronc, branches, feuilles, aidant à satisfaire à la demande croissante de pâte à papier.

19. See page 61 — This Morbark Chipharvester machine handles the whole tree — main stem, limbs, and foliage, helping to meet the growing demand for fibre for pulp.

Kapitel 9/Chapitre 9/Chapter 9 — Zeiten/Pages/Pages 89 — 93

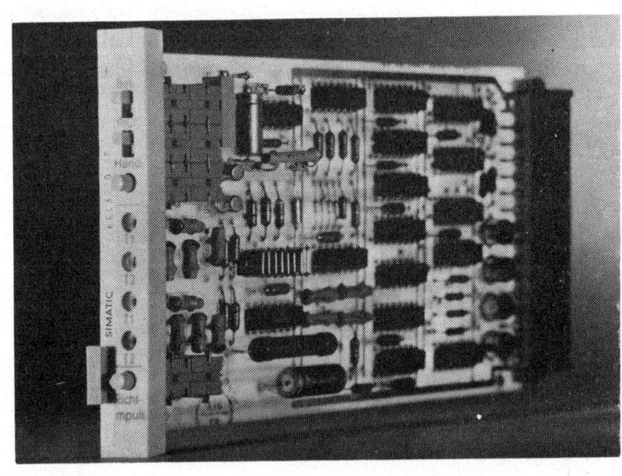

20. Seite 99 — Zur Verarbeitung der Messdaten und Signale stehen Standardbaugruppen aus dem SIMATIC C3 - System zur Verfügung. Dieses System arbeitet mit integrierten Schaltkreisen.

20. Page 103 — Blocs standard du système SIMATIC C i utilisé pour mesurage des grûmes et commande des machines. Les blocs ont des circuits intégrés.

20. See page 91 — Standard modules from the SIMATIC C3 System are used to process log measurement data and machine commands. Modules have integrated circuits.

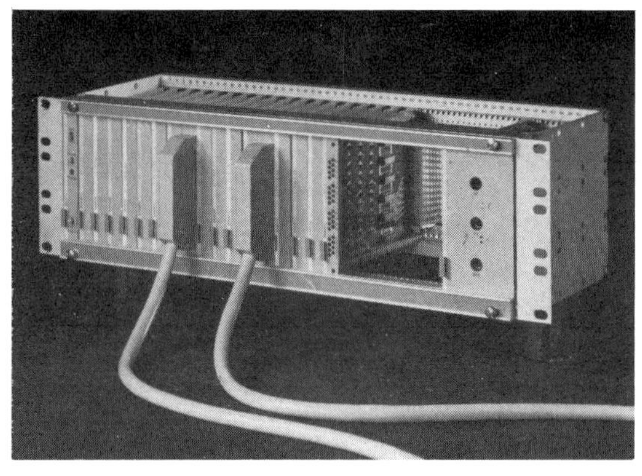

21. Seite 99 – Eine Anzahl von Steckbau-
gruppen werden in zugehörige Baugruppen-
träger untergebracht und dort zu Funktions-
einheiten verdrahtet.

21. Page 103 – Un certain nombre de blocs
peuvent être montés en châssis et branchés
pour obtenir des unités fonctionnelles.

21. See page 91 – A number of modules
can be mounted in a cabinet rack and connected up to form functional units.

Kapitel 11 / Chapitre 11 / Chapter 11 – Zeiten/Pages/Pages 118 – 131

22. Seite 127 – Das patentierte Führungs-
system der Canadian Car (Pacific) gewährt
grössere Sägensteife und dynamische Be-
feuchtung. Es ist ein Spiralaxiallager mit
flüssigem Schmiermittel.

22. Page 131 – Système de guide de
Canadian Car (Pacific) assure rigidité et
amortissement dynamique. Basé sur le prin-
cipe du palier de butée hélicoïdal avec un
peu de lubrifiant.

22. Page 118 – Guide system for circular
saws, operating as a spiral thrust bearing
increases saw stiffness and provides dynamic
dampening. Added liquid ads lubrication.